VOLCANO WEATHER

VOLCANO WEATHER

THE STORY OF 1816, THE YEAR WITHOUT A SUMMER

Henry Stommel and Elizabeth Stommel

SEVEN SEAS PRESS
NEWPORT, R.I.

PUBLISHED BY Seven Seas Press, Inc., Newport, Rhode Island 02840

Edited by James R. Gilbert

Portions of this book were previously printed
in *Scientific American.*

FIRST PRINTING, June 1983
SECOND PRINTING, August 1983

3 5 7 FF/FF 0 8 6 4 2

LIBRARY OF CONGRESS CATALOGING IN PUBLICATION DATA

Stommel, Henry M., 1920–
    Volcano weather.

    Includes bibliographical references and index.
    1. Weather—Effect of volcanoes on—Case studies.
    2. Tambora, Mount (Indonesia)—Eruption, 1815.
    I. Stommel, Elizabeth.   II. Title.

    QC981.8.V65S86   1983   551.6   82-19658

    ISBN 0-915160-71-4

Designed by Irving Perkins Associates
Printed in the United States of America
by Fairfield Graphics

Dedicated to the memory of

PRISCILLA BRAISLIN MONTGOMERY
(1874–1956)

and the
Marine Biological Laboratory Library,
which she served so well

CONTENTS

vii

CHAPTER 8

FLEEING THE SCRABBLE FARMS

Westward migration 92

Crowding of the land • The forces behind emigration • The case of Vermont • Demographic data • Comparing 1816 with the Oklahoma Dust Bowl • Patterns of flight • Land booms • Efforts to stem the tide

CHAPTER 9

TALL TALES AND EXAGGERATIONS

Legend and folklore 101

Legends as folk history • The man who froze to death • The man who lost his toes • Embroideries upon fact • Grandfather Hopkins • Suicide • Slaughter of sheep • Semi-starvation • How the Old Farmer's Almanac got an undeserved reputation for foresight

CHAPTER 10

THE CHOLERA CONNECTION

A distant link to 1816? 109

What cholera is like • Confined to Asia before 1816 • Carried by British troops • Moslem pilgrims infected • Reaches the Caspian shores • Tsar Alexander's efforts to confine it • Poles accuse Russians of biological warfare • Paris can't bury its dead • The French prime minister succumbs • 50,000 die at Mecca • New York City quarantined, 100 die each day.

CHAPTER 11

OF LIGHTNING RODS AND SUNSPOTS

Does anybody really know what causes climate change? 116

Alarms over sunspots in 1816 • Electrical heating of the Earth • Iceberg outbreak hypothesized • Deviations of the Gulf Stream discounted • Benjamin Franklin's speculations • Efforts to document the weather • Early temperature measurements • Development of thermometers and temperature scales • Tree rings and coral bands • Incompatibility of old and new data • The volcano hypothesis • Recent temperature tabulations • Trends of past 200 years • Difficulty of interpretation • The "little ice age" • Climate and sunspots • Random fluctuations • The "dust veil index" • Computer simulations • The 1963 eruption of Mount Agung • Representativeness of old data • Records in Greenland ice • Mount St. Helens and the weather • The 1982 eruption of El Chichón

PREFACE

IN THE SPRING of 1982 a great cloud of volcanic dust and acid droplets in the upper atmosphere spread around the world. It was attributed to the eruption of Mount Chichón in Mexico. The amount of sunlight reaching the earth's surface was reduced five percent. Climatologists speculated that the temperature of the world might be lowered for a year or two.

New Englanders found the summer of 1982 strangely cold and wet. They shivered at the beach, bewailed their faltering gardens and wondered at the August snow in the mountains of Vermont. Newspapers pictured skiers gliding down the slopes of Mount Killington. A shortfall in the California tomato crop was predicted. The National Oceanic and Atmospheric Administration set up a special task force to monitor what was happening.

The full story of the possible effects on climate from this great volcanic cloud will not be known until several years have passed and all the world's data is assembled and studied. Meanwhile, television commentators and newspaper columnists speculate on possible changes in climate. And they remind their viewers and readers about the year 1816, when New England and Western Europe suffered a year without a summer, when crops were short, hunger threatened the remoter hill farms of New England and Canada and actual famine stalked regions of Europe.

The purpose of this book is to relate, so far as we are able, what actually happened in the period 1815–1817 and to try to make a balanced judgment concerning the probable links between the cataclysmic eruption of Mount Tambora and the weather that followed. It is a kind of retrospective "impact statement." For those who demand certainty, it will exhibit most of the weaknesses of such statements: Too little is known about these vastly complex physical and social phenomena to be sure about causes. However, we can discover a good deal about what happened in 1816. It is a story worth telling.

Cape Cod, September 1982

VOLCANO WEATHER

CHAPTER 1

THE VOLCANO THAT COOLED THE WORLD

Mount Tambora erupts

IT WAS THE spring of 1815. There was intense excitement in Europe. Napoleon had escaped from the prison island of Elba, landed at Cannes on the French Riviera on the first of March and, gathering admirers about him, proceeded in triumph to Paris. Louis XVIII had fled the city and the heady Hundred Days, which were to end in Waterloo, had begun.

On the other side of the world, the islands of the Dutch East Indies lay quiet in their humid heat. The British, under Sir Thomas Stamford Raffles, with the help of the resident Dutch, had driven out Napoleon's appointee Hermann Willem Daendels. The long string of southern Indonesian islands, stretching from Sumatra on the west to New Guinea on the east, slumbered peacefully in the equatorial sun. In the middle of this island chain is the island of Sumbawa, and on its northern shore stands the great volcano Mount Tambora. This mountain was about to explode in an eruption which, according to some modern meteorologists, sent more dust into the upper air and obscured sunlight more than any other volcano between 1600 and the present. It would be the most explosive eruption in the last 10,000 years.

The island's tranquility was first disturbed on the evening of April 5 by a series of deep shocks. At a distance they sounded

Thanks to Sir Thomas Raffles, temporary lieutenant governor of the Dutch East Indies, modern scientists have an accurate record of the eruption of Mount Tambora in 1815— the largest volcanic eruption in the last 10,000 years. COURTESY NATIONAL PORTRAIT GALLERY, LONDON, ENGLAND.

very much like cannon firing. At Djogjokarta (some 450 miles west of Mount Tambora) a detachment of soldiers was sent out to determine whether some nearby military posts were under fire, or perhaps if a ship along the coast was signalling an emergency. The government vessel *Benares* was in the harbor at Macassar, the capital of the southern Celebes. Those on board also heard what they thought was a cannonade somewhere to the south. Towards sunset, the fusillade seemed nearer. The *Benares* was dispatched with some Dutch troops to inspect islands to the south. After three days' fruitless search for what he thought might be pirates, the captain of the *Benares* returned to Macassar. When the explosions began again on the evening of April 11 and houses and ships began to shake, he again set sail southward to see what was happening. At dawn on the 12th, he wrote in his log that the sky was very dark. By noon the heavens were so black and the air filled with so much fine ash that it appeared a realization of Milton's "darkness visible."

It was phenomena such as these that alerted the inhabitants of Sumbawa's neighboring islands that the great Tambora volcano was in the throes of an enormous eruption. No native records of the eruption survive. It is only from accounts written by a few of the thin sprinkling of European colonizers that any chronicle of events exists to describe what surely was a vast human tragedy.

The British survey the damage

As temporary lieutenant governor of Java, Raffles requested British residents of the islands to report to him on the effects of the eruption. He dispatched an assistant, Lieutenant Owen Phillips, to make a general survey of the plight of native villages. Raffles presented a summary of these reports to the Natural History Society in Batavia in September, 1815. From this account, Sir Charles Lyell excerpted the following description of the eruption in his epoch-making book, *Principles of Geology* (1830), a book that had an impact upon Mosaic view of geolog-

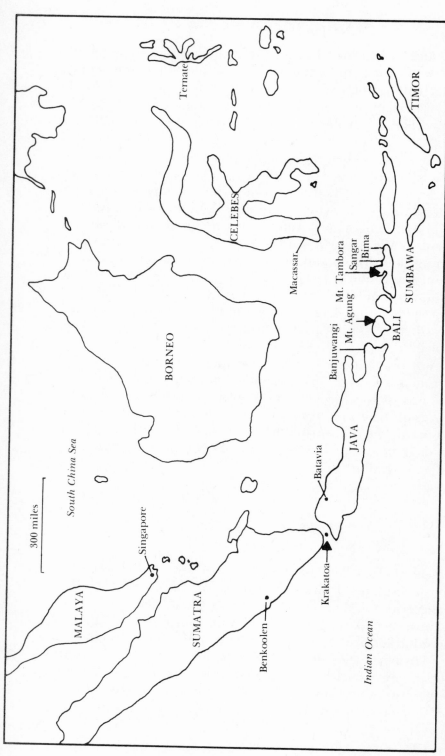

Map showing the location of three great volcanos—Krakatoa, Mount Agung and Mount Tambora.

ical history similar to that of Darwin's *Origin of the Species* upon biology some decades later:

> "Island of Sumbawa, 1815—In April, 1815, one of the most frightful eruptions recorded in history occurred in the mountain Tambora, in the island of Sumbawa. It began on the 5th day of April, and was most violent on the 11th and 12th, and did not entirely cease till July. The sound of the explosions was heard in Sumatra, at the distance of nine hundred and seventy geographical miles in a direct line, and at Ternate in an opposite direction, at the distance of seven hundred and twenty miles.
>
> "Out of a population of twelve thousand, only twenty-six individuals survived on the island. Violent whirlwinds carried up men, horses, cattle, and whatever else came within their influence, into the air, tore up the largest trees by the roots, and covered the whole sea with floating timber. Great tracts of land were covered by lava, several streams of which, issuing from the crater of the Tambora mountain, reached the sea.
>
> "So heavy was the fall of ashes, that they broke into the Resident's house at Bima, forty miles east of the volcano, and rendered it, as well as many other dwellings in the town, uninhabitable. On the side of Java, the ashes were carried to the distance of three hundred miles, and two hundred and seventeen towards Celebes, in sufficient quantity to darken the air. The floating cinders to the westward of Sumatra formed, on the 12th of April, a mass two feet thick and several miles in extent, through which ships with difficulty forced their way.
>
> "The darkness occasioned in the daytime by the ashes in Java was so profound, that nothing equal to it was ever witnessed in the darkest night. Although this volcanic dust, when it fell, was an impalpable powder, it was of considerable weight: when compressed, a pint of it weighing twelve ounces and three quarters. Along the sea-coast of Sumbawa, and the adjacent isles, the sea rose suddenly to the height of from two to three feet, a great wave rushing up the estuaries, and then suddenly subsiding. Although the wind at Bima was still during the whole time, the sea rolled in upon the shore, and filled the lower parts of the houses with water a foot deep. Every boat was forced from the anchorage and driven on shore.

Batavia, seat of colonial government in the Dutch East Indies, as it appeared circa 1780 in this aquatint by Ivan C. Rynne. In April of 1815, Batavia was covered with a half-inch of ash in the aftermath of Tambora's explosive eruption more than 700 miles away.
ILLUSTRATION COUTRESY E. T. ARCHIVE, LONDON, ENGLAND.

Satellite view of Mount Tambora.

A 1944 U.S. Air Force reconnaissance photo of Mount Tambora, showing view from the north. Note the large lake at the bottom of the crater.

"The area over which tremulous noises and other volcanic effects extended was one thousand English miles in circumference, including the whole of the Molucca Islands, Java, a considerable portion of Celebes, Sumatra and Borneo. In the island of Amboyna, in the same month and year, the ground opened, threw out water, and then closed again."

The sound of the distant detonations of Mount Tambora were taken to be signs that the gods of the mountains were about to free the islands from European rule—so the priests of Java proclaimed. In the old Dutch trading port of Grissee, north of Surabaya and 300 miles west of Tambora, a white

SOME MAJOR ERUPTIONS
FOR COMPARISON

Year	Name	Country	Explosive Rank	Volume Ejected (cubic km.)
1470 B.C.	Santorini	Greece	2	10
79 A.D.	Vesuvius	Italy	3	1
1815	*Tambora*	Indonesia	1	100
1883	Krakatoa	Indonesia	2	10
1963	Agung	Indonesia	4	0.1
1980	St. Helens	U.S.A.	3	1

Our "EXPLOSIVE RANK" = 8 –(Volcanic Explosivity Index)

Santorini's explosion gave rise to the ancient legend of Atlantis. Vesuvius buried Pompeii. Tambora, however, ranks as the most explosive volcano in the past 10,000 years, according to a ranking by the Smithsonian Institution. Krakatoa was the first volcano subjected to serious scientific study and is known to have reduced direct sunlight by 15 to 20 percent. Agung is a modern volcano whose meteorological effects are well documented. Mount St. Helens seemed to have no marked meteorological effects at all.

trader who rode about the countryside after the major eruption of April 13 was told that the goddess Loroh Kidul was
about to have one of her children married and the eruptions
were in honor of the event.

Later, when a general survey of the thickness of the volcanic
ash that had been deposited could be made, it was found to be
three feet at Sangar, one and a half feet at Bima—some 14
hours by road from Tambora—one foot on the island of Bali,
nine inches at Banjuwangi and about the thickness of a dinner
knife at Batavia. According to Lieutenant Phillips, houses in
Sangar and Bima were half-buried by the ash and, of course,
the harvest covered up. The ash also seemed to cause a severe
diarrhea—a plague that affected horses and cattle as well as
people. The ensuing famine was severe.

Mount Tambora ejected some 25 cubic miles of debris, reducing the mountain some 4,200 feet in height. Dust was
thrown into the high stratosphere, where it would circle the
globe for several years and, perhaps, reduce incoming sunlight
by reflecting some of the sun's light back into space. Most of
the debris fell back on land as a layer of pumice a foot thick—
even hundreds of miles away. Large islands of floating pumice
were seen at sea and were encountered by ships up to four
years later.

*A section view of Mount Tambora (seen NE to SW) and Mount St. Helens,
drawn in scale, showing the comparative loss of mountaintop during their
eruptions.*

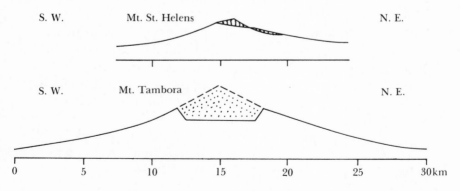

The catastrophic outburst and accompanying earthquake claimed 90,000 lives. There was total darkness by day for three days within 200 miles of the volcano. According to the most recent (1981) listing of the Smithsonian Institution, Mount Tambora's explosion ranks first out of the 5,564 dated eruptions, with an estimated volume of ejected ash about 100 times that of Mount St. Helen's 1980 eruption. It exceeded the more famous and better documented explosion of Krakatoa in 1883, and even that immense paroxysm amongst the Grecian Islands of 1470 B.C. that gave birth to the legend of Atlantis. Mount Tambora's dust cloud, which encircled the earth for several years, has been cited as the cause of the cold year that followed.

Ascents to the summit

The first scientist to visit Mount Tambora following the eruption was a biologist, Heinrich Zollinger, who climbed to the rim of the crater from the east in 1847—a time within living memory of the eruption. He recorded that before the eruption the mountain was a striking cone approximately 4,000 meters high, the loftiest point in the East Indies and well wooded. This peak was reduced by the eruption to a jagged-edged caldera whose rim did not exceed 2,950 meters in altitude and which was nearly circular with a diameter of six kilometers. The floor of the crater was depressed some 600 to 700 meters below the level of the rim. Zollinger spent only an hour at the rim, but left the first useful sketch map.

The second known ascent of Mount Tambora was by a University of Basle geologist, Pannekoek van Rheden, who reached the rim from the southern side on September 14, 1913. He made a second map, obtained some photographs and geological specimens, but again did not spend more than a few hours at the summit.

Between 1928 and 1930, a famous oceanographic expedition sailed to the Dutch East Indies. The colonial government had ordered a new hydrographic surveying vessel, the *Willebrord Snellius*. It was arranged that before it was delivered officially to the government for hydrographic use, it would first be made

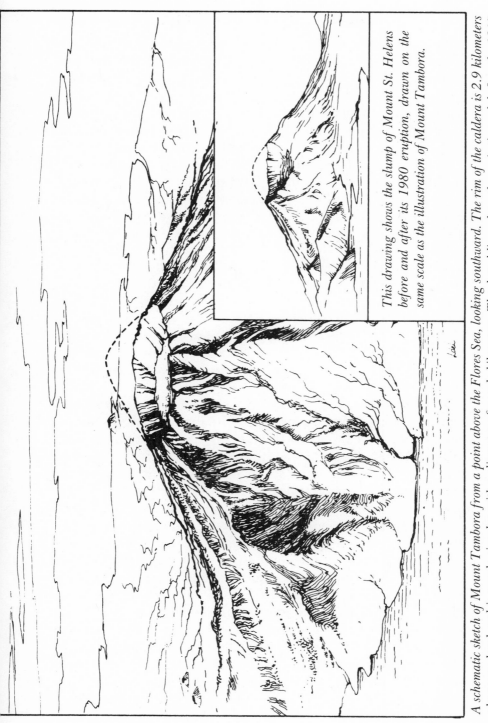

A schematic sketch of Mount Tambora from a point above the Flores Sea, looking southward. The rim of the caldera is 2.9 kilometers above sea level and is nearly circular with a diameter of nearly kilometers. The dotted lines show the majestic peak before the 1815 eruption. The crater today is 600 meters deep.

This drawing shows the slump of Mount St. Helens before and after its 1980 eruption, drawn on the same scale as the illustration of Mount Tambora.

available to a Mr. P. M. van Riel for a privately financed oceanographic study of the archipelago. With his wife's money, the
topside chartrooms were converted into comfortable living
quarters and a suitable number of native servants were en-
gaged. The results of this expedition—which must have been
one of the most pleasant of all time—were excellent. Amongst
these were extensive sediment samples from the ocean bottom,
which provided material for determining the distribution of
the ash from the 1815 Tambora eruption.

Between 1933 and 1947 several ascents were made—for the
sport of it—from the west side. One was made by government
officials and another by the manager of the Tambora Coffee
Estate, Mr. G. Bjorklund. Another ascent, by geologist W. A.
Petroeschevsky, was made in 1947. Petroeschevsky landed at
the western end of the Sangar Peninsula from a sailing proa at
the fishing village of Kananga. He then walked up about 12
kilometers to the Tambora Coffee Estate, where he spent the
night. Beyond the estate the old trail was so overgrown it had
to be cut open again by plantation laborers. Four whole days
were required to reach the last camp, about 500 meters below
the steep walls of the rim.

Petroeschevsky visited the rim for an hour that evening and
again the next morning for four hours before descending. He
produced a sketch map of the topography of the inside of the
crater. So far as can be determined from scientific literature
and from correspondence with volcanologists, no one has yet
climbed down into the caldera. However, Petroeschevsky has
given a fascinating description of it as seen from the rim, of its
grasses, sulphur deposits, small craters and vents, and of a
freshwater lake fed by rains with a grove of trees beside it.

Indonesia became independent in 1949. Since then, studies
of its volcanos have been conducted under the auspices of the
new republic's Department of Volcanology. Expeditions to the
rim were made in 1976 and 1981, the former in collaboration
with the geology department of the University of Tasmania.
Volcanic eruptions are common in Indonesia, although no
eruption before or since was as great as that of Mount Tambora. They are still regarded with religious awe.

In early 1963 Bali's Mount Agung provided climatologists

A 1724 drawing of Ternate, another Indonesian volcano.

Panoramic view of the crater of Mount Tambora, shot from the western rim in 1981. COURTESY INDONESIAN VOLCANOLOGICAL SURVEY.

Two more views, shot in 1976, of Mount Tambora's massive, 1,800-feet-deep crater. PHOTOS COURTESY OF DR. J. D. FODEN.

with an excellent test of the idea that the dust from volcanic eruptions can lower global temperature. We will refer to its scientific importance later in Chapters 11 and 12.

The Agung eruption occurred while worshippers were praying on the mountain at the great Temple of Besakih, participating in the celebration of the Great Rite of the Eleven Directions. The government's Department of Foreign Affairs officially explained the eruption as a sign from the mountain gods that "men have not yet succeeded in restoring the balance of nature in purification and renewal which bring harmony and happiness to mankind." It also was alleged that the date of the ceremony—which is held only once a century—had been chosen incorrectly. The pamphlets of the Tourist Department had already been printed, however, and the governor ordered the Great Rite to proceed as planned, despite the risk of affronting the mountain deities.

Rich offerings were made on the slopes of Agung, in the presence of the high gods. Cattle, pigs and poultry—their necks weighted with stones—were blessed by flower-decked priests and then dropped into a nearby lake to please the water spirits. The villages were crowded, and multitudes at the time of the eruption were on the mountain. A total of 1,550 people

The lake in Mount Tambora's crater may be a relatively recent geographical feature. It was first noted in W. A. Petroeschevsky's 1947 ascent of Tambora.

lost their lives, 85,000 were made homeless and a third of the farmland on Bali covered with ash and lava.

After the blast, relief supplies were sent by Canada, the United States, the Philippines, Thailand and France. Many natives, however, refused the relief and chose to suffer without help, to expiate their guilt and the displeasure of the gods.

CHAPTER 2

"EIGHTEEN HUNDRED AND FROZE TO DEATH"*

The cold summer in New England

TURNING NOW FROM the Tambora eruption to the climatic events that followed, it is fair to ask what science actually knows about the cold summer of 1816. The answer, surprisingly, is quite a lot. Just as it was fortunate to have had an educated English governor with a budding interest in natural science in Indonesia—and anxious to record events there—it is also fortunate to have had in New England a group of men who had the leisure and interest to record the weather through the years before and after 1816. And, most important for science, they had adequate meteorological instruments—particularly good thermometers.

President Edward Holyoke of Harvard College in Cambridge, Massachusetts, was keeping a meteorological temperature record as early as 1749, using a poorly calibrated thermometer given to him by the Royal Society of London.

Hanging in an unheated room of his house, it used the antiquated Hauksbee temperature scale (0°H at body heat, 65°H at freezing). Better instruments were obtained at Mannheim University by Dr. Samuel Williams, Harvard's Hollis Professor of Natural Philosophy, and a more accurate series begun. But, unfortunately, Dr. Williams was detected embezzling the col-

*The wording "Eighteen hundred and froze to death" is a genuine colloquial expression commonly found in historical literature about the summer of 1816.

19

Harvard College as it appeared in the early 19th century.

Impressive Yale College amidst sleepy early 19th century New Haven.

A clergyman and president of Yale College, Jeremiah Day was busy writing books on elementary mathematics, navigation and surveying in 1816, the year he recorded New England's unusually cold weather. COURTESY YALE LIBRARY.

lege's Hopkins Trust Fund and he had to flee to Vermont to escape criminal indictment.

Due to Dr. Williams' moral lapses, the Cambridge temperature record was not reestablished until July, 1816, by Dr. Benjamin Waterhouse, the superintendent of New England's Army Hospitals and the physician who introduced smallpox vaccination to America.

Good temperature records were being kept at other colleges, including at Middlebury, Bowdoin and Williams. The longest continuous weather record in New England was at Yale College in New Haven. Yale's record, continuous from 1779, when an earlier record by President Ezra Stiles was interrupted by a British invasion and the thermometer broken, was maintained faithfully by successive Yale presidents, clergymen-scholars with regular habits willing to rise at 4:30 a.m. to read the instruments.

During the summer of 1816, the Yale observer was Professor Jeremiah Day, who had turned from the ministry to an academic career because of a throat hemorrhage brought on by too-strenuous preaching. When the average of his June, 1816, temperature readings is compared to other Junes, it can readily be seen how cold June of 1816 really was: about 7°F below normal—a temperature that in normal Junes is found only some 200 miles north of Quebec City. June, 1816, was by far the coldest June ever recorded at New Haven.

Another impressive record was being maintained in New England's leading commercial port, Salem, by the town's physician, Dr. Edward Augustus Holyoke, son of the Harvard president. Begun in 1786 and running to 36 manuscript volumes by the time of his death at more than 100 years of age, it contained four daily observations of the thermometer, barometer and general condition of the sky.

Farther south along the coast the little 300-house village of New Bedford was not yet, in the words of Herman Melville, a city of fine houses "harpooned from the world ocean," with long avenues of green and gold maples upon the corners of which one might encounter "actual cannibals chatting" and "women blooming like fine roses." But it already was alive with maritime activity and a whale fishery second only to that of Nantucket. At his home on the northwest corner of Water Street and William Street, a serious young Quaker merchant,

Mean June temperatures at New Haven, Connecticut, for some early years.

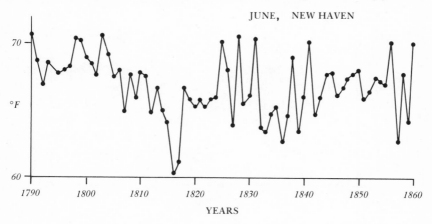

JUNE, NEW HAVEN

YEARS

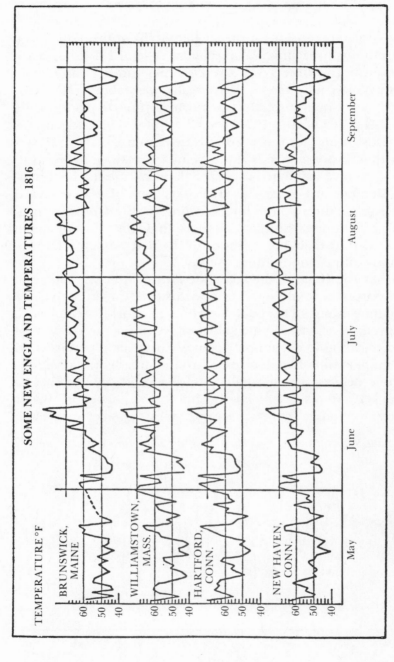

Some New England temperatures—1816

Samuel Rodman, Jr., already had begun his meteorological journal in 1812, a record that was destined to be kept up by his descendents until 1905.

In addition to these meteorological records, there were diaries, letters and newspaper reports.

June, 1816, began auspiciously enough considering the backwardness of that year's spring. Crops that had survived mid-May frosts and lack of rain were beginning to show progress at last. But on June 6 the first of three unseasonable cold waves crossed into New England from Lake Champlain and, moving eastward, had covered all of New England by the end of the day. The cold and wind lasted until June 11. In northern New England the storm left three to six inches of snow on the ground. A second killing frost struck the same areas on July 9 and a third and fourth on August 21 and 30, just as the summer's ravaged harvest was about to begin. Freezing weather destroyed all but the hardiest grains and vegetables.

Indian corn, the New England staple crop, was killed back severely, despite attempts to replant it. Most of the summer was also very dry, which compounded the farmers' difficulties. Farmers felt the effects of the cold weather differently according to their locality. The effects were most severe for farmers in the far north and more moderate for those along the coast.

The southern limit of the heavy snows of that early June lay close to Bennington, Vermont, a land, in the eyes of Reverend Timothy Dwight, "whose soil was of the first quality, equally suited to all the productions of the climate." Dwight saw that "wheat and grass, the extreme of agricultural production, grew there luxuriantly and alike, that the pastures were covered with rich and abundant herbage."

Hiram Harwood was a young farmer in 1816, whose father was the first white child born in Bennington. Harwood was of a literary inclination and kept an unusually informative diary. During the last days of May, he remarked upon the "hindrance of vegetation" by frosts earlier that month, which he estimated to be about two weeks late.

On the last day of May, he spent the morning planting pumpkin seeds, setting traps for crows and cleaning the house drain. The afternoon was warm and he walked with his father, who was not feeling well, on the mountain to gather their flock

Farmer Benjamin Harwood, father of diarist Hiram, in an old oil painting. PHOTO COURTESY THE BENNINGTON MUSEUM.

The Harwood family farm as it appeared in 1899 in Bennington, Vermont. COURTESY THE BENNINGTON MUSEUM.

of sheep for marking and driving to Wilmington some score of miles to the east along roads that must today be Route 9.

Snow in June

When he went to bed on the evening of June 5th, Harwood's head was full of thoughts about a wonderful story he had heard the night before about a barrel of human bones dug up by doctors, told him by Mr. Parsons who had just returned from a visit to Hoosack Four Corners. Harwood did not dream of the great storm racing toward Bennington across western New York State and Pennsylvania. It rained much during the night, the wind blowing sharp gusts from the northeast. By morning, it had started to snow.

When he awoke early on the 6th, he saw that "the heads of mountains on every side were crowned with snow—the most gloomy and extraordinary weather" he had ever seen. He spent the morning weighing wool and counting fleeces. He counted 38 and knew that five or six of his flock were still at large. Accordingly, his father went sheep hunting during the afternoon to the west end of town but with no success.

On the morning of June 7, Harwood found the surface of his ploughed fields stiff with frost. The leaves of the trees were now blackened. A washtub full of rain water was skimmed with ice. Snow remained on nearby Sandgate and Manchester mountains until past noon. The wind was extremely strong and the cold abated but little in the afternoon. His father and a neighbor rode till noon hunting the missing sheep. Harwood himself mended fences on the north side of his wheat field with greatcoat and mittens on, at which task his father joined him in the afternoon.

On Saturday, June 8, the awful scene continued with sweeping blasts from the north early in the day spawning intermittent snow squalls. During the morning, while digging stone in the sheep pasture, it was so cold that Harwood and his father were "absolutely compelled" to send for their mittens and to wear them till midday. Towards nightfall, when he and his cousin Sam decided to weed the garden, they were actually frozen out. The next day, Sunday, was frosty but perfectly clear and Harwood enjoyed the peace and satisfaction of his reading and

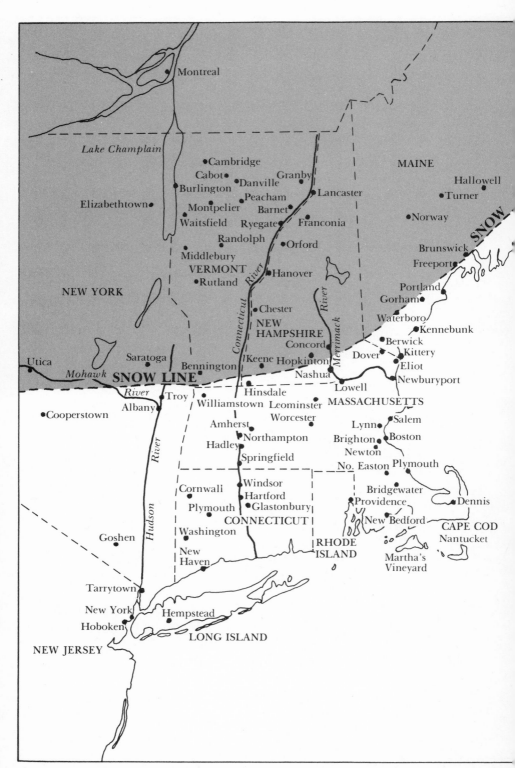

Snow fell as far south as western Massachusetts in June of 1816. In southern New

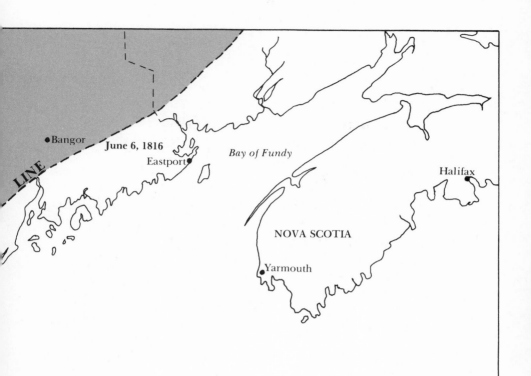

●Bangor   **June 6, 1816**

Eastport●

Bay of Fundy

LINE

Halifax
●

NOVA SCOTIA

●Yarmouth

Atlantic Ocean

England, farmers experienced sleet, snowsqualls and crop-killing frosts.

writing. Monday, June 10, was also a cold day and he wrote that he was obliged to thrash his hands while hoeing. The corn, which had been up a few days before, was "badly killed and was difficult to see." Three of his sheep were found in the eastern part of town.

In New Hampshire, Governor William Plumer kept a careful temperature record (1796 to 1823) but it suffers because he moved about a good deal, his weather journal accompanying him from his farm in Epping to the capital in Concord and elsewhere on his travels about the state. Thursday, June 6, was the day set for Governor Plumer's inauguration address. The ceremony was held at the meeting house. The address is memorable because it contained a proposal to revise the charter of Dartmouth College, effectively putting it under state control— and the governor succeeded in stirring a legal hornet's nest over the sanctity of charters, a battle that went all the way to the supreme court. Governor Plumer's journal refers to the backwardness of the season. On the morning of June 7, he recorded a low of 38°F, and on the 8th more snow falling.

It so happens that another account of the weather at the inauguration survives, perhaps a little embellished by many tellings, but still with a ring of truth. It is the reminiscences of Sarah Anna Emery, whose grandparents were in Concord that morning in the company of their friends, Mr. and Mrs. Daniel Colman. On the way to the meetinghouse to attend the inauguration, the "wind blew a gale with an occasional shower of snow flakes—it was so powerful that it was difficult to reach church even with the gentlemen's assistance." Once seated, they strove to enjoy the scene, but in vain. "Our teeth chattered in our heads, and our feet and hands were benumbed. Mrs. Colman had a troublesome tooth and it begun to ache unbearably."

Dinner at the hotel was scarcely better. The wind was so powerful at Concord Bridge, she wrote, they "thought the chaise must be overturned; even Mr. Emery, who never feared anything, was a little discomposed." The night was passed at a hotel on the Chester Turnpike, where in company with other travellers they shivered around a rousing fire, complaining of the cold room.

In Plymouth, Connecticut, the apprentice clockmaker

A 1793 engraving of Dartmouth College.

The June 15, 1816 edition of the Danville North Star *described the "melancholy weather" New Englanders were experiencing.*

NORTH STAR.

DANVILLE, *Saturday*, June 8, 1816.

THE WEATHER.

COMPENSATION *to the Members of Congress.*

REPUBLICAN PRINCIPLES
Of the Federal Party.

SPANISH AMERICA.

FEDERAL APATHY.

MASONIC.

GOODS,

Hard Ware,
Crockery & Glass Ware,

GROTON.

STRAY MARE.

Chauncey Jerome was setting up some machinery at the Terry Clock Factory. According to his memory of the events, recorded years later in an autobiography, the scene was quite unforgettable.

> "I well remember the 7th of June, while on my way to work, about a mile from home, dressed throughout with thick woolen clothes and an overcoat on, my hands got so cold that I was obliged to lay down my tools and put on a pair of mittens which I had in my pocket. It snowed about an hour that day. On the 10th of June, my wife brought in some clothes that had been spread on the ground the night before, which were frozen stiff as in winter."

Most New England newspapers commented on the unexpected cold wave. On June 8, 1816, the *North Star* at Danville, Vermont, remarked:

> "*The Weather*—Some mention was made not long since of the unusual backwardness of the spring, and the remarkable instability of the weather. Although the summer months have commenced, the weather is no more steady nor the prospects more promising. Wednesday (June 5) last was perhaps as warm and sultry a day as we have had since September—at night heat lightning was observed, but on Thursday morning the change of weather was so great that a fire was not only comfortable, but actually necessary. The wind during the whole day was as piercing and cold as it usually is the first of November and April. Snow and hail began to fall about ten o'clock, A.M. and the storm continued till evening, accompanied with a brisk wind, which rendered the habiliments of winter necessary for the comfort of those exposed to it. The snow nearly all melted as it fell—sufficient remained, however, on Friday morning, where the wind still continued piercing, to give the ground the appearance of winter, and to afford a gloomy prospect of the progress of vegetation, the cold the preceding night having been so severe, that the ground was considerably frozen, and water, in some instances, froze nearly half an inch thick."

This was followed in the June 15, 1816, issue by:

> "*Melancholy weather*—Some account was given in last week's issue of the unparalleled severity of the weather. It

continued without any essential amelioration, from the 6th to the 10th instant—freezing as hard five nights in succession as it usually does in December. On the night of the 6th, water froze an inch thick—and on the night of the 7th and the morning of the 8th, a kind of sleet or exceeding cold snow fell, attended with high wind, which measured in places where it was drifted, 18 or 20 inches in depth. Saturday morning the weather was more severe than it generally is during the storms of winter. It was indeed a gloomy and tedious period . . . Wednesday, partly cloudy, but oppressively hot and sultry. Thursday, cold and squally, with a sprinkling of rain and snow, which was quite visible on the trees of Ascutney Mountain. Friday clear, except flying clouds, and so cold, especially in the forenoon as to require almost winter fires. Saturday, still uncomfortably cold, squally and blustering—winter fires, and winter groups around them. Considerable snow has fallen in this state and N. Hampshire. Probably no one living in the country ever witnessed such weather, especially of so long continuance!"

The cold wave also was noticed much further to the south. According to the *New Bedford Mercury* of June 14: "The weather during the latter part of last week was very cold, winds high and frost severe—most of the garden vegetables and etc. were destroyed. The Catskill Mts. on the 6th inst., were covered with snow. In the northern parts of Delaware County, the crops of corn are mostly cut off."

Snow did not fall in New York City, but warm-weather birds did. The *Commercial Advertiser* reported that great numbers of birds of all kinds generally only found in distant forests were driven to the city and dropped dead in the streets.

On the fateful Thursday, June 6, 1816, the Connecticut diarist Reverend Thomas Robbins narrowly missed getting caught in the storm, having just returned to his home in South Windsor from a meeting of the Hartford North Association in West Hartford. His diary notes that the weather was becoming quite cold and windy. Robbins, scion of a long line of preachers, after some five years of missionary work in Vermont and the Western Reserve—his health broken—was serving as pastor in South Windsor. Like other pastors at that time who eked much of their living from the soil, Robbins was a parttime farmer,

Portrait of Thomas Robbins, COURTESY THE CONNECTICUT HISTORICAL SOCIETY.

although a fairly bookish one, having acquired a library of some 600 books. His agricultural efforts, detailed in his diary, showed he was attentive to the influence of weather on the progress of crops. Just the year before he had published a study entitled: *"An Historical View of the First Planters of New England."*

During the June cold wave Robbins required a steady fire and became "oppressed with anxiety." By Sunday, June 9, he was so much concerned about the continuing cold and wind and the injury to fruit trees and gardens that he preached to his congregation on the parable of the Fruitless Fig Tree (Luke XIII). He recorded frost the next two days and supposed the corn to have been "killed to the ground."

In North Branford, Connecticut, Calvin Mansfield entered in his Memoranda of the Seasons for June 11: "Great frost—we must learn to be humble."

From killing frost in Connecticut to snow in Quebec City deep enough to reach the axletrees of the carriages, the weather was, in the words of Adino Brackett of Lancaster, New Hampshire, "beyond anything of the kind I have ever known."

Until June 11, New Englanders shivered and got out heavy clothing. Fireplaces and stoves were rekindled. Gardens were blackened by the frost. Sheep that had been shorn perished, even those which were housed.

In the wake of the storm good growing weather returned. From Connecticut to Quebec farmers with reserves of seed ploughed their blackened fields and replanted their corn, beans and other tender crops.

Frost in July

The second cold wave came in early July. But it was not so severe. Chauncey Jerome in Plymouth, Connecticut, remembers seeing "several men pitching quoits in the middle of the day with thick overcoats on, and the sun shining bright at the time."

On July 5, in Maine, ice froze as "thick as window glass," and on July 9, corn was again killed by frost except in very sheltered places. At Warren, Maine, when corn was being hoed for the first time, the frost cut it down again. Professor Parker Cleaveland at Bowdoin College in Brunswick, Maine, found his thermometer standing at 33.5°F on the morning of July 9. Frost was reported in places nearby. The nights were cool and unfavorable for vigorous growth of corn.

	Thermometer				Barometer			Rain	Wind		
1	50	75	66	63.7	30.15	30.05	30.00		SW	W	W
2	64	75	69	67.7	29.95	29.85	29.82		S	S	SW
3	66	71	58	65	29.80	29.90	29.95		N	N	NW
4	48	69	60	59	30.00	29.93	29.80		NW	S	S
5	62	70	66	69	29.65	29.50	29.50	.71	SW	W	W
6	55	52	39	48.7	29.55	29.65	29.70		W	NW	NW
7	35	57½	41	44.3	29.72	29.70	29.70		NW	NW	NW
8	38	56	42	45.3	29.75	29.80	29.90		NW	NW	NW
9	39	65	50	51	30.00	30.02	30.03		NW	NW	N
10	39	60	59	49.3	30.07	30.10	30.17		N	N	SW
11	36	60	51	49	30.20	30.20	30.15		N	S	NW
12	48	67	52	55.3	30.00	29.92	29.85		SW	SW	SW
13	50	75	62	62.3	29.86	29.85	29.82		SW	SW	SW
14	60	68	65	64	29.82	29.80	29.90	1.02	N	S	S
15	61	68	62	63.7	29.90	30.00	30.02		E	E	E
16	54	66	61	60.3	30.03	30.03	30.03		NE	SE	E
17	51	64	54	56.3	30.05	30.04	29.95	.25	NE	N	NE
18	53	60	61	58	29.90	29.90	29.90	.36	N	N	NE

Meteorological journal for New Haven kept at Yale College for many years reflects the unusually cold weather that began on June 6, 1816, and continued through June 11. The upper part of two pages face each other in the journal.

In Vermont, Middlebury College's Professor Frederick Hall recorded a temperature of 34°F at 7 A.M. on July 8 with frost —the unrecorded temperature presumably being lower earlier in the morning. Calvin Mansfield of New Branford, Connecticut, recorded a "cold, small frost" the same day. On July 8, the frost was so severe at Franconia, New Hampshire, that it killed off all the beans. On July 10, there was frost in the lowland at Chester, New Hampshire.

The cold was still more severe in Canada. The small lakes to the north of Bay St. Paul on the St. Lawrence River were still covered with ice in the middle of July. In places, the ice was strong enough to bear the weight of Indians.

The corn crop failure was now certain in New England. But

1816

Weather

1	clear	clear	clear	
2	clear	clear	clear	
3	br cloud	hazy	clear	
4	clear	br cloud	br cloud	
5	br cloud	clear	rain	
6	br cloud	br cloud	clear	Gale on the southern coast of U.S.
7	clear	clear	clear	Frost
8	clear	clear	clear	Frost Snow in Vermont
9	clear	clear	clear	
10	clear	clear	clear	
11	clear	clear	clear	Frost
12	br cloud	clear	clear	
13	br cloud	cloudy	cloudy	Snow in Maine
14	rain	hazy	br cloud	Thunder shower in the morning
15	hazy	br cloud	clear	
16	cloudy	br cloud	clear	
17	cloudy	br cloud	rain	

it did not seem to awaken much concern among the newspapers of the coastal towns. In July there certainly was no real prospect of famine. Only an occasional short note, as that appearing in the Portland, Maine, *Argus* of July 17 advised farmers to plough their grasses under again and to resow them so that new growth could be expected by the end of September.

The following notice was carried in the July 23 issue of New Hampshire's *Dover Sun:* "To Farmers: It is acknowledged on all hands that the first crop of grass has been very light; perhaps not more than half the usual quantity. To make up for this deficiency it is recommended to farmers to plow down as much ground as convenient as soon as possible and broadcast with oats and Indian corn. These will be fit to cut about the 30th of September when the saccharine juices of the corn-blade and stock together with the tender oat straw of the oats will make fodder equal to the best hay—try and be convinced."

Further, responding to the concern for finding food for cattle during the coming winter, the *New Bedford Mercury* carried the following advice: *"Substitutes for Hay*—A New Hampshire paper says, an excellent fodder for horned cattle may be collected from potatoe tops. It is a practise in many places at the Southward to reap off about two thirds of the length of their potatoe tops, and dry them on mowing land in the usual way of hay making. Several tons may be collected from an acre, and no damage to the potatoes, if taken as soon as they are ripe, and before the leaf begins much to fall."

The chief concern was that farmers might find it difficult to feed their animals, not that they would be hungry themselves.

The latter half of July was good warm growing weather, although it was rather dry. The early part of August also seemed like true summertime. Working in his warm Connecticut garden, the Reverend Robbins felt his health improve. He was happy with thoughts of a wonderful change seemingly taking place in public sentiment. "Almost all the newspapers now publish religious intelligence," he noted.

Frost again in August

As July turned into August, and the heat persisted, it became apparent that there also was a drought.

By August 17 Robbins was busy watering trees and the following night held a season of prayer on account of the drought at the meetinghouse. This was immediately answered with a shower—unfortunately a trifling one. On August 20 there was morning frost, but the drought persisted. Towards evening of the 25th he got down to business and held "a solemn and interesting season of prayer" for rain, which was promptly answered next morning with a very refreshing shower.

Sidney Perley, in his *Historic Storms of New England* describes the weather of August, 1816, in the following way:

> "To the twentieth of the month the weather had been warm and pleasant, but on that day squalls occurred in New Hampshire, rain falling at Chester, and snow on the mountains at Goffstown. At Keene, the change in temperature on that day was greater than had ever been observed in this variable climate. On the night of the twenty-first, there was

a frost, which at Keene and at Chester, New Hampshire, killed a large part of the corn, potatoes, beans and vines, and also injured many crops in Maine. It was felt as far south as Boston and Middlesex county in eastern Massachusetts, and in the western portion of the state as far as Stockbridge, where it injured vegetation. The mountains in Vermont were now covered with snow, and the atmosphere on the plains was unusually cold.

"In Keene, New Hampshire, the oldest persons then living said that they never saw such a severe frost in August. It put an end to the hopes of many farmers of ripening their corn, especially in the low lands, and they immediately cut the whole stalks up for fodder, but being in the milk it heated in the shocks and spoiled. By the twenty-ninth of the month the frost had reached as far south as Berkshire County, Mass., where it killed the Indian corn in many of the fields in the low lands. The farmers there saved much of it by cutting it up at the roots and placing it in an upright position, where it ripened upon the juices of the stalks. If frost had kept off two weeks longer there would have been a very good crop of corn in Massachusetts."

On August 20 Governor Plumer was travelling from Concord to Hanover, New Hampshire. He found no prospect of an Indian corn crop from Salisbury to Hanover. On August 21 he noted "a hard frost, that in many places of vast extent killed Indian corn (particularly in pine lands), potatoe vines, pumpkins, cucumbers, etc. We shall have but a small crop of corn—that which is not killed is chilled. On Connecticut River the night fog prevents early frost." On the 30th he returned from Hanover to Concord and wrote that the prospect of a corn harvest was small but that grain of every other kind was abundant. "Apples few till I reached Salisbury [where they] were thick. Hay about two thirds in quantity and three fourths in quality of a common crop. People busy gathering it in."

If the corn harvest was bad, there seem to have been compensations of nature. The Dartmouth, New Hampshire, *Gazette* for September 18, 1816, said, "the fruit harvest is plentiful, apple and pear trees are shored to prevent the limbs breaking under the weight of their fruit."

A good and attentive farmer, such as Bennington's Hiram

Harwood, did not suffer overmuch. He mowed hay during the first weeks of August and began to cradle his spring wheat. Tuesday, August 20, was windy and cold. On August 21 some people saw frost. By the 23rd, Farmer Harwood's winter wheat was also cradled and carted in. The next day he commenced on the oats.

On August 25, suffering from toothache—and even though it was Sunday—Harwood took up some more oats. The following day it rained early and, in company with his father, he reaped "as fine a crop of oats as is rarely seen."

White frost hit again on August 29, and he finished carting 12 loads of 10 bushels each of winter wheat (a total of 3,000 sheaves) from the 14-acre field. As the month ended Harwood had harvested in addition to the winter wheat, 244 sheaves of spring wheat, 920 of rye, 810 of oats, three cartloads of barley and two of flax.

To emphasize that the summer of 1816 was extraordinary even when compared with the other summers of the generally cold 1811 to 1820 decade, examine the tabulation made by Alfred Loomis of the dates of last frost of spring and first frost of fall in New Haven for that decade. In 1816 the last frost of spring is given as June 11, 20 days later than the corresponding frosts for the rest of the decade. The first frost of autumn is given as August 22, 35 days earlier than the mean of the other years of the decade. Thus the frost-free period of 1816 in New Haven was 55 days shorter than its usual span of 126 days—and this does not even take into account the frost recorded in early July.

At the end of the summer, and into the fall, no one seems to have suspected a tie-in with the eruption of Mount Tambora. Far away as it was from New England, American sea captains were roaming the Pacific and often visited Batavia. Observations of large amounts of pumice floating in the Indonesian seas following eruptions have always been common. The mariners of Salem, Stonington and New York were not strangers to these seas. But considering that at the time the Dutch East Indies—and any news from those parts—were half a year away from New England, it is understandable that nobody connected New England's aberrant weather with an antipodean volcano.

Observations of a nearby iceberg, however, were taken as newsworthy and relevant. Thus, when Captain Gooday of the ship *Jones* encountered a mile-long iceberg near the Grand Banks on August 31, it was mentioned in the *New York Gazette*, presumably because it would interest those already thinking about the lost summer.

CHAPTER 3

FAMINE IN EUROPE

Legacy of a cold, wet summer

BY MID-SEPTEMBER in 1816, when New Englanders had a fair estimate of the extent of their crop shortages, news of an even more disastrous summer in Europe began to appear in newspapers from correspondents abroad.

The Albany *Argus* of July 22, 1816, carried a letter from London that reported: "From the Baltic to Breslau—the greater part of the land sown with winter grain has been obliged to be ploughed up and that of the corn that remains standing scarcely one third part of a crop is to be expected.

"All the accounts from Germany, the Netherlands and Switzerland, agree in stating that so deplorable a season was never known in the memory of man." Readers were informed that in Germany a proposal had been made to prohibit distillation of spirits from corn in order to protect grain supplies for making bread and providing seed for the next year. Letters from France warned that the vintage would be bad.

The Albany *Argus* of January 21, 1817, reported from Ireland: "A famine in this unfortunate country is inevitable for the harvests entirely failed from the badness of the weather." The prediction proved itself: Ireland's famine led to a typhus epidemic that in the years 1817 to 1819 afflicted 1,500,000 people and killed 65,000. The typhus spread all over Europe. Public health facilities were heavily burdened. For example, by late 1817 Edinburgh's Royal Infirmary was filled to overflowing

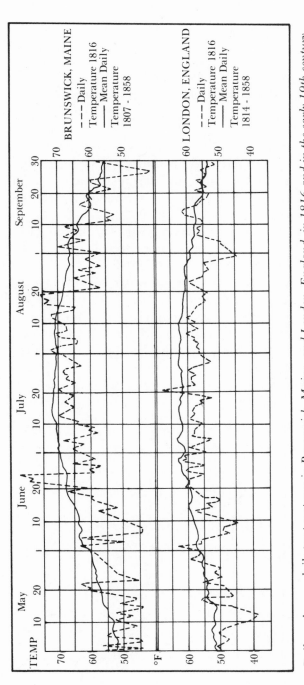

Daily and mean daily temperatures in Brunswick, Maine, and London, England, in 1816 and in the early 19th century.

and the Queensberry House Barracks was commandeered to serve as a temporary fever hospital.

The cold summer struck just as Europe was emerging from the chaos of the Napoleonic Wars. George III was still the king of England in the 56th year of his reign, Louis XVIII was king of France, in the third year of the Bourbon Restoration and Napoleon had entered exile in the South Atlantic island of St. Helena. Lord Byron, the poet, had run away from his wife in London, and had taken the Villa Diodati on the shores of Lake Geneva as a refuge. He and his house guests were kept indoors by the cold rainy weather, and wrote ghost stories to amuse themselves. It was there that Mary Shelley wrote her famous novel *Frankenstein*.

The cold summer of 1816 set many records in Europe. July of 1816 was the coldest July in the English Lancashire plain in 192 years of record. According to the long records obtained at Geneva the mean summer temperature of 1816 was the lowest since 1753. In an 1883 study of the dates of wine harvests in France and Switzerland between 1782 to 1856, Charles Alfred Angot showed that in Switzerland, 1816 was the year of the latest harvest date for Lausanne, Le Crest, Aubonne, Vevey,

Dates of the wine harvest in Lausanne, Switzerland, showing the late harvest of 1816.

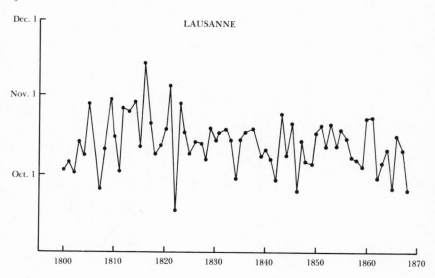

and in France for Medoc, Tain, Volnay, Dijon, Bourges, Vendome and Vesoul. Other districts show 1816 simply as a very late harvest. In Beaune three other harvests were later. In Couvignon, 1817 was just as late as 1816—all pointing to local deviations from a perfect score.

Switzerland in distress

Switzerland was distressed by famine. Since the time of the middle ages, Zurich had been a center for the grain market. The prices chronicle periods of extreme scarcity, including 1692 and 1770–1771 and, equally, the high prices of 1816–1817. Local authorities record "the year 1816 was more than usually dry, cold, and unpleasant." Attempts to replant summer wheat were frustrated by lack of seed in the state granaries. Swine had to be slaughtered due to lack of fodder. Potatoes were not universally accepted as fit for humans (in Tosstal a student refused to sit down next to a fellow student because he ate potatoes)—but nonetheless many were forced to eat them. By August, 1816, even the supply of potatoes had run out. More than 30,000 persons were jobless and breadless. But it was in the industrialized, heavily populated regions that the distress was most horrible.

By mid-1817 the price of grain in Switzerland had tripled. The painter Rudolph Tanner published an aquatint showing those who died of famine watching gourmandizers being led to Hell. Pastors wrote desperately for help. Parish records of Ueti-kon, for example, register deaths with clear symptoms of starvation. On January 26, 1817, churches declared a day for special collections to alleviate distress. All kinds of things were eaten: sorrel, Iceland moss and cat flesh. Instructions were issued to help identify poisonous plants. In Constance, a Dr. Sauter published directions for making ersatz bread from wild plants. Zurich was overwhelmed by adult and child street beggers, despite strenuous efforts of plainclothes police to drive them out.

At the height of the famine, Swiss cantons sealed themselves off, prohibiting the export of grain to one another—a most

This aquatint, by Rudolf Tanner, shows Swiss gourmandizers in the famine of 1817 about to be led to hell. PHOTO COURTESY KUNSTHAUS, ZURICH.

unusual domestic measure in this ancient confederation. Trading houses like Gebr. Finsler and Salomon Pestaloz sought far and wide to purchase grain. They purchased supplies in Lombardy, and in Venice they were able to buy Odessa wheat. But frequently the shipments were intercepted by hungry bandits in the passes and near Lake Como.

The Zurich city government allocated special funds to alleviate the famine. Smaller towns did the same. For example, in 1817 Wadenswil, a town of 4,000 souls, provided its 800 poor with 8,000 gulden in cash, 1,200 for soup, 530 for meal, 170 for bread and 100 for potatoes. The soup, incidentally, was made according to the recipe of Benjamin Thompson of Woburn, Massachusetts. Barley, peas and potatoes were boiled for three hours in water and served in bowls to which rye bread and vinegar were added.

Memories of the famine of 1816–1817 were invoked by the Swiss over the next 20 to 30 years as justification for accelerated experimentation and agricultural research.

Riots in France

The situation in France was equally grave. Torn by military campaigns and the defeat at Waterloo in 1815, and stripped of the feudal protections that had been one of the positive aspects of the vanished aristocracy, the French peasantry had no reserves to buffer the bad season of 1816. While the Royalists led by Louis XVIII and Tallyrand struggled to maintain the constitutional monarchy against the followers of Napoleon who sought a more liberal form of government, France seemed to be on the verge of a replay of the revolution.

Political ferment was reflected in the slowing down of industrial activity. This, in combination with the meager harvest of 1816 and the high price of available food—nearly exhausted by the invasions of 1815—set the stage for riots and insurrection.

In Poitiers, rioting broke out because of a 3 franc-a-bushel tax imposed on wheat. Grain carts on their way to market in towns along the Loire Valley had to be protected by soldiers and gendarmes, who found themselves fighting on occasion as many as 2,000 hungry and enraged citizens. Where the harvest had been good, farmers feared to take their produce to market

Medallion struck in Southern Germany in memory of the great famine of 1816–1817. The inscription reads: "Great is the distress, Oh Lord, have pity." COURTESY IRENE BURKHARDT, PHOTO WAYNE MCCALL.

in the face of so many robbers and brigands—including some-times the national guard itself.

Where authorities intervened and threatened to dictate the availability of food—or imposed a tax upon it, as in Toulouse —the public was quick to resent this as a form of paternalism suggesting a return to prerevolutionary government.

By December, 1816, so many food-related crimes were com-mitted each day that the authorities in Montpellier, in the south of France, no longer took the trouble to seek out the guilty. As the winter progressed with an ever-worsening shortage of food, the thievery grew. On January 17, 1817, at the market of Fauville, near Yvetot, only 50 sacks of grain were available in-stead of the usual 800 to 900. Anticipating a riot, a detachment of soldiers was dispatched. Crying, "down with bayonets," the crowd quickly overwhelmed them with stones, sacked the city hall and took what grain there was. This scene was reenacted in little towns all over France, the grain insurrection "growing like a fire."

Where the harvest was adequate, as in the western regions— with perhaps some small reserve—the people were infuriated by transportation of grain to regions where the shortages were most severe. Often local officials took part in the riots, feeling that the harvest of their region belonged to its inhabitants and exportation was a violation of their security. In Normandy, where shortages were extreme, the high prices together with taxes amounting to 24 francs a hectolitre (2.89 bushels) were usually paid, but not without protest. As early as August 7, 1816, the government suspended import duties on grain. And on November 20 France began to import grain and subsidized the markets in Paris to the amount of 30 million francs.

According to J. P. Housel's recent study of the mean monthly national prices of wheat from 1801 to 1912 the very highest point was reached in France during the first half of 1817, ap-proximately twice that of the long-term average, a factor rather similar to that in the United States.

Violence and disorder continued unabated through the spring and early summer of 1817 until the harvest of that year brought some amelioration of Europe's food-supply troubles.

In the words of Northeastern University Professor John D.

Post—who has analyzed the economic disorders in Europe at the time of this famine—the years 1816–17 were the "last great subsistence crisis in the western world."

Other regions

In Japan, where meticulous records of rice yields were kept, 1816 does not appear to have been exceptional. In China the summer, though cold, was not extraordinarily so. Keeping weather journals was as common amongst the Chinese leisure class as bird-watching is among us today. Chinese meteorological records of the time are therefore very numerous, but unfortunately do not include thermometer readings, and hence it is impossible to make firm comparisons of the summer of 1816 with other summers. It would appear, however, that the truly exceptional character of 1816 weather was limited to a small portion of northeastern America, Canada and the extreme western parts of Europe.

CHAPTER 4

LIFE ON THE SCRABBLE FARM

Climate's effect in an era
of self-sufficiency

To OBTAIN A balanced view of the summer of 1816's impact on the economy of New England it is necessary to know what the economy was like. How did most people make their living? How prepared were they for the trials and privation 1816 brought? In the absence of help from the government, how did they survive?

In Europe, as we have seen, the consequences were more severe—even chaotic—and governments did intervene. But extensive famine did not stalk the United States as it did Europe.

In 1816 the youthful nation was recovering from the War of 1812, a war from which the United States gained no territory, no redress for the impressment of seamen, no fishing rights. In fact the Treaty of Ghent settled none of the issues for which the war was originally declared. Even the nation's pride over its greatest military victory, the Battle of New Orleans, was tainted because it was actually waged after the peace treaty had been signed.

A new sense of responsibility grew within the federal government toward the country as a whole. The divisiveness over the unpopular war, which had almost led to the secession of New England states attending the Hartford Convention, was quickly forgotten. The country surged forward on a new wave of nationalism. But with a federal tax revenue of $3 a year per person, the federal government was in no position yet to offer extensive help in time of economic emergency.

The population of the entire United States was approximately that of New York City today. The center of population lay in the Shenandoah Valley. Only Pennsylvania, New York and Virginia had more than a million people. But of the five other states with populations over half a million, Kentucky and Ohio were evidence of the great westward migration and settlement that were even then underway. The populous state of Kentucky had been admitted to the Union in 1792, Ohio in 1803. New admissions came in rapid succession in the decade 1810 to 1820: Louisiana (1812), Indiana (1816), Mississippi (1817), Illinois (1818) and Maine (1820).

The inland economy at the beginning of the early 19th century was nearly static, not having changed much in the past 50 years. It would soon change with the advent of industry and manufacturing, but at the time there was a pause in "progress." The early days of pioneering were long past in Connecticut, Rhode Island and Massachusetts, but still very much alive in northern New England.

The state of the roads

The most important factor constricting the growth of the interior was the absence of convenient, inexpensive transportation. Along the coast, most shipments were by sea. Surplus garden produce from Connecticut, for example, moved to New York City in coasters. The principal rivers also provided access to and from inland farming regions. The Hudson was a pathway to market for the tidy Dutch truck farms along its bank; the Connecticut breathed life into Hartford. New engineering works, like the South Hadley Canal, were opening access even further upstream. In fact the canal's novelty was great enough to entice the Reverend Robbins to inspect it during his August visit to his friend Reverend Samuel Osgood in Northampton.

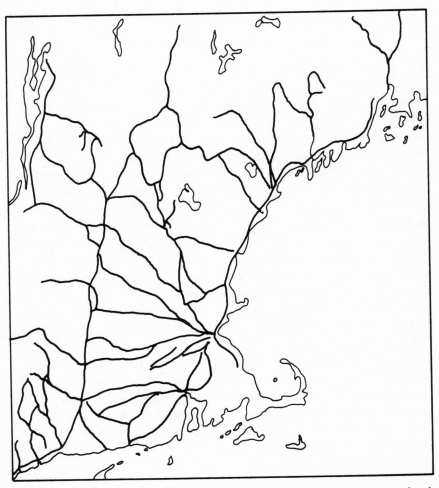

Public roads and private turnpikes were few and far between in New England in the early 1800s. Coastal transport was by sea. By stage coach Boston to Providence: 9 hours; Boston to Albany: 4 days.

The Merrimac River was connected to Boston markets by the recently completed 27-mile Middlesex Canal. There were even steamboats on some of the rivers. Robert Fulton's fleet of Hudson River steamers numbered six by February of 1815 when he died of exposure after rescuing a lawyer friend who fell through the ice at the Jersey City ferry landing.

But a few miles away from the banks of these rivers was sufficient an obstacle to prevent transporting such bulk products as corn and cordwood to city markets. Only agricultural products in concentrated, processed form—such as pearlash (collected from burning trees while clearing the land), cheese, cider and whiskey—could be profitably transported more than 15 miles by land.

Today the northeast states depend upon transportation of foodstuffs from distant western and southern states by rail and truck and even by air. In 1816 there were no railroads. There were perhaps 4,000 freight-carrying wagons in the whole country. Roads were circuitous and ill-kept and offered very limited access to the back country and hilly regions. Bulk cargos could be transported economically only by water. This meant that inland towns and farms were very much on their own.

A very good picture of the difficulty of inland travel and transportation in New York and New England is to be found in Yale President Timothy Dwight's published travels. Over a period of 20 years (1795–1815) between college terms, accompanied by a son and a friend or professor of the college, he traveled throughout New England. His personal impressions of the people, the state of the church, the local history of settlements and towns and a conscientious description of the countryside were carefully preserved in his journal. From his accounts it is possible to picture in great detail the landscape and the life of New England and New York in the early 19th century.

The very first roads beyond the environs of the cities and commercial ports began as "mere passages from farm to farm." As a township developed, the necessity for broader communication caused these to be joined with one another by the simplest of horse paths, reaching in time to the larger centers of commerce. In hilly or mountainous country in Vermont, western Massachusetts and New York the obstacles presented by rocks and forest, swamp and river were such that it was often necessary to travel far out of one's way to reach a settlement. Dwight, for example, traveling 14 miles through Dalton to Lancaster, New Hampshire, found the road "so absolutely covered with stones as to have given rise to a proverbial remark that in this spot no horse ever set his foot on the ground."

As the inhabitants of any particular settlement were responsible for the building and improvement of local roads and bridges, these were accomplished slowly and imperfectly to the meagerest specifications. As Dwight informs us: "The inhabitants . . . are so few, so poor, and so much occupied in subduing their farms and in providing sustenance for their families that it is often a long time before bridges are repaired. Such, upon the whole, was the state in which we found them that they soon became objects of more dread than any other inconvenience attending our journey."

Though the state legislatures appropriated certain sums of money from time to time for the upkeep of the post roads, the matter of road building was still a haphazard and disorganized undertaking in 1816. In a journey made from Providence to Plainfield, Connecticut, Dwight notes, "The people of Providence expended upon this road . . . the whole sum permitted by the legislature. This was sufficient to make only those parts which I have mentioned. The turnpike company then applied to the legislature for leave to expend such an additional sum as would complete the work. The legislature refused. The principal reason for the refusal, as alleged by one of the members, it is said was the following: that turnpikes and the establishment of religious worship had their origin in Great Britain, the government of which was a monarchy, and the inhabitants slaves; that the people of Massachusetts and Connecticut were obliged by law to support ministers and pay the fare of turnpikes, and were therefore slaves also; that, if they chose to be slaves, they undoubtedly had a right to their choice, but that freeborn Rhode Islanders ought never to submit to be priest-ridden, nor to pay for the privilege of traveling on the highway. This demonstrative reasoning prevailed, and the road continued in the state which I have mentioned until the year 1805. It was then completed, and freeborn Rhode Islanders bowed their necks to the slavery of traveling on a good road."

By contrast, Dwight is quick to recognize the efforts of local residents where they succeed, as in a trip through towns bordering on Lake Winnepesaukee: "It ought to be observed to the honor of the inhabitants of this state (New Hampshire) that, although the population is sparse, they are making their roads universally very good. In the parts where they were originally the worst, they have already made them to a great extent excellent, in the manner of turnpikes, and better than some roads that wear this name."

The word turnpike here is of course a term used to designate a road large enough to carry a horse and carriage but carries

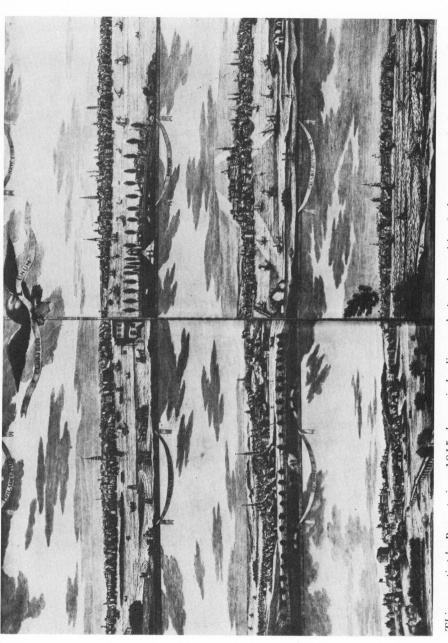

This aquatint by Boquet circa 1815 shows six leading early American cities. It demonstrates the importance of water —as the only reliable form of transportation—to the economic development of young America.

no guarantee of a road passable in all kinds of weather. Without benefit of surfacing material the roads were subject to washouts and gullys caused by heavy rain. Fallen trees might lie where they fell until the local residents had time and inclination to saw them up and remove them. Here, for example, Dwight describes a journey taken through Goshen, New York, to the head of Lake George, where "Instead of the smooth way which we expected, we found one more irregular and embarrassing than any which either of us remembered . . . The darkness was so profound that our horses were unable to grope out their course, and became fearful and hesitating. At times they refused to advance; and at times, with a trembling, tripping step they appeared ready to plunge us in the mire. With a continual alternation of ascents and descents between the path, the ditches, and the drains, we found our way entangled and distressing . . . with a creeping, snail-like pace."

Though night journeys were seldom undertaken by plan, the slowness of travel often precipitated unsoughtfor eventide adventures: "The darkness had become intense and palpable: the branches of the trees on the opposite sides of the road meeting over our heads and excluding the faint light of the stars, so that we were obliged to trust ourselves wholly to the guidance of our horses. A wolf, which I presume considered us as having sufficiently lost our wits to become his lawful prey, howled after us at a small distance; but perceived that we were too numerous to be attacked . . . In the night companies of wolves have compelled individuals traveling alone to betake themselves to trees for safety, and confined them to this unpleasant lodging until morning."

In 1815, Lewis Downing of New Hampshire produced his famous Concord Coach, the first such contraption to make possible a comfortable ride. In ensuing years he built 3,000 of these "Stage-coaches," which opened the west to travel.

The stability of a subsistence economy

Isolated as they were, accustomed to the privations caused by subsistence farming, 90 percent of the population of New England was essentially self sufficient. A further strength of the

New England farmer was that he owned his farmstead outright by fee simple. There was no mortgage to meet, no banker to threaten and the minute taxes could be met by a few days' labor for the town. The land was his, unfettered by encumbrances of any kind, cleared by his own arm or that of his father. This was in marked contrast to the helpless state of the tenant who was later to be driven from his farm in the great depopulation of the Oklahoma Dust Bowl during the Great Depression.

In the back country, without public utilities or access to markets, farmers were quite capable of surviving for a season on shortened rations. They knew how to improvise. They could search the woods for wild fruits and ate the inner bark of trees when there was nothing else. Nesting in the woods were wild fowl and deer for meat and, if there was a stream nearby, there was fish as well.

A farm of 100 to 200 acres provided nearly everything that a family needed to survive. Six acres of apple orchard assured an adequate supply of cider and 40 acres of woodland provided timber for building and cordwood for burning.

Less than an acre would be set aside for flax, sown for the manufacture of linseed oil and linen. But this was one of the smaller and less successful crops. There were several acres devoted to raising corn and rye, the essential grains cultivated by every farmer in New England. The average corn crop of about 28 bushels per acre supplied meal, mixed one third corn with two thirds rye flour to make the everyday "rye and Injun bread," without which no family could have survived.

Pumpkins and potatoes along with small amounts of oats and barley were grown for animal fodder, though they were used in the farm kitchen as well. A little buckwheat was sown to clean the fields of weeds and to supply flowers for the bees. Wheat was early abandoned as a crop in most of New England, except the Connecticut River Valley, because of a ruinous disease known as "blast" or "smut"—a kind of mildew.

The kitchen garden was much like our vegetable plot of today. The *New Bedford Mercury* advertises in its April 26, 1816, issue seeds for white, yellow and red onion, beets, carrots, crook neck squash, salmon and scarlet radish, early curled lettuce, "Yorkshire cabbage," peas, beans as well as melons and a variety of herbs. The emphasis was on vegetables that could be stored without special attention over the cold winter. This contributed to an inevitable monotony of diet through most of the year.

Berries and grapes that grew wild were eaten when ripe and, when there was a sufficient supply of maple or West Indies sugar, were made into preserves. The apple was a most versatile and valued fruit. Made into apple sauce, apple butter and apple pie, apples were also converted to "apple-molasses" and to cider, the everyday drink. Apple slices strung on linen threads were hung to dry in attics, available for use when the barrel of applesauce in the cellar was gone.

It took several cattle to supply enough milk for cheese. Like the swine, they roamed the cornfields for gleanings and depended on what nuts and roots they could find. Seldom fed

In 1801 Seth Adams of Massachusetts smuggled a pair of Merino sheep out of Spain. Next year, General David Humphreys, the U.S. minister to Spain, brought over an entire flock. The Merino sleep craze had begun. The wool was extolled, so much so that one of Humphrey's rams sold for $1,500, the equivalent of a year's wages for a sailing ship master.

grain, they suffered from exposure in winter and from poor breeding.

Sheep survived the long, harsh winters in somewhat better condition and 1816 was a year of heightened interest in sheep raising: Merino sheep wool sold at $1.50 a pound, marking the beginning of the New England "sheep craze." The newly imported Merino sheep were found mostly on the prosperous farm. A ram was often worth $1,000. But pork in its various forms was the meat the New England farmer depended upon most.

Each family was an economic unit, though neighbors depended upon one another for mutual assistance in house-raising, haying and harvesting, grinding of corn, weaving of cloth and the manufacture of shoes and hats. It was a usual thing to exchange goods and services—a joint of meat for the grinding of corn at a grist mill, or the exchange of a tanned hide for the dying and weaving of wool. An occasional piece of furniture was bought when there was sufficient cash to pay for it. So treasured were their dishes and cooking pots that each of these items might be accounted for in wills.

The few things for which cash had to be found were sewing needles, cotton thread, a pair of eyeglasses or an iron cooking pot, all purchased from itinerant peddlers. In towns large enough to have a general store there was a greater opportunity at the same time to bring produce or articles made at home to market. A 19th century Barnet, Vermont, couple could claim on their 60th wedding anniversary never to have bought any meat, flour or sugar at the store during their married life.

In the farm family of 1816 there was a natural division of tasks, the men assuming the heavy work of shearing the sheep and washing the wool; or breaking, swingling and hackling the flax to prepare the fibers for spinning. Many a backwoods farmer wore the same best suit of itchy homespun material from his wedding day to his funeral.

While all the children went to school long enough to learn to read, write and "cypher," they would be soon needed at home for help in the multitude of tasks associated with their self-sufficiency.

Without a hired man—and few farmers could afford one

Every early 19th century farmer in New England raised hogs. And nearly every meal included some pork product.

even at a wage of 60 cents a day—the farmer was hard pressed to keep up with the demands of what today would be considered a large family farm.

Careless about manuring, weeding and rotating crops and limited in his knowledge of farming, the Yankee farmer aimed to supply the needs of his own family. Such a simple means of enriching the soil as applying manure was usually neglected through ignorance. It was jokingly said that some farmers would build a new barn rather than distribute to the fields the accumulated manure in an old one.

Without his woodlot the inland farmer could not have survived. Wood was the sole source of heat and cooking fuel. While saltwater farmers on Cape Cod were beginning to experiment by burning peat in their stoves, farmers elsewhere were burning huge amounts of wood in their fireplaces. Nearly all farm implements, plough and harrow, rake, hoe and shovel, were made of wood and made by the farmer himself, often reinforced or edged with bands of iron. All household furniture, carts and carriages, kitchen bowls, spinning wheels and looms were made of wood. The ingenious farmer sometimes made these things for market if he was near enough to reach one with oxen and cart.

The constant drain on woodland by such multifarious needs had, by 1810, already used up most first-growth timber and few farmers knew anything about woodland management. Wholesale cutting led to a shortage of firewood. In 1816 the dwindling of the woodpile in summertime must have led to melancholy thoughts.

Coal was not used for fuel, even in the cities, until in later years it could be moved by canal and railroad.

The farmer's biggest problem came in the spring of 1817, when there was no seed for new crops and no money with which to buy it. Efforts to assure the availability of seed corn were made by private individuals and by local governments.

Thus, the May 18, 1817, edition of Portland, Maine's *Eastern Argus,* records a Town Meeting authorizing "the Overseers of the Poor to furnish seed of various descriptions to those individuals who are unable to procure the same from his own resources—the advances to be paid for either in labor on the

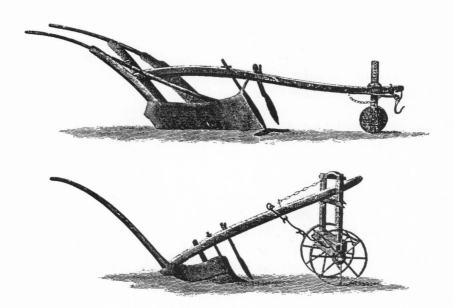

Tilling New England's rocky soil was no mean feat. Many different plows were developed to meet the challenge, as shown in these illustrations from an 1817 edition of the Massachusetts Agricultural Repository and Journal, COURTESY THE AMERICAN ANTIQUARIAN SOCIETY, WORCESTER, MASSA-CHUSETTS.

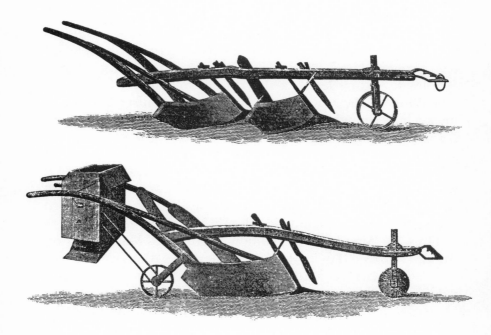

highway, or in kind at the harvesting of the crops. We sincerely hope that this example may be generally followed in other towns."

Tight fisted, ornery and independent—perhaps a little like his Appalachian cousins today—the Yankee farmer gained a reputation as a rather sloppy jack-of-all-trades. His independence was largely a matter of necessity and it would stand him in good stead when he was faced with the scanty harvests of 1816.

Indians guarding the corn fields as drawn by Capt. S. Eastman.

CHAPTER 5

————————————————

THE REAL DISASTER
IN NEW ENGLAND

Failure of the corn crop

IN THE SUMMER of 1816 corn ripened so poorly that not more than a quarter of what was sown in Connecticut was usable for meal. The rest—unripe, mouldy and soft—was fed to hogs and cattle but did little to fatten them.

Although Indian corn was the only major casualty of the cold summer of 1816, it was no by means a minor matter. Those who think of it as a late-summer luxury that graces dinner tables for a few short weeks have little appreciation of its importance. It was *the* staple crop of rural 19th century New England.

It was first seen on November 5, 1492, by Columbus while exploring the Caribbean. He saw fields 18 miles long.

In Virginia the corn fields expanded rapidly, from a mere 30 acres in 1609, to 500 in 1614. By 1631 Virginian planters were exporting corn to Europe.

In Massachusetts, corn assured the survival of the Pilgrim settlers. On November 15, 1620, only a few days after their first landing at what is now Provincetown on Cape Cod, they were being urged by the impatient captain of the *Mayflower* to disembark. Captain Miles Standish, with a band of 16 men, spied six Indians and a dog walking along the beach.

Upon approaching them, the Indians fled away—not altogether unnaturally because they had massacred the crew of a wrecked French vessel just three years before. Standish and his party attempted to follow them for a parley. They lost their quarry, but found instead a field abandoned by the Indians, which included some sand hills padded down by hand. Upon opening these caches they found baskets of corn, some on the ear. They took some of the ears to the ship, which soon thereafter sailed on to Plymouth. It was from this seed that the Pilgrims established their first corn plantations in old Indian fields. The Indians taught them how to plant it with a fish in each hill for fertilizer.

For the early New Englanders corn was the best crop. The seed were already acclimated, it served both man and animal, and it was the most adaptable and best-yielding crop for newly cleared land.

Corn today is not so very different from what it was in Columbus's time. Sweet corn was known to the Iroquois and Missouri tribes, but was not known to the settlers until 1779, when Lieutenant Richard Bagnall, on Sullivan's Indian Expedition, returned from a campaign west of the Susquehanna with several sugary-kerneled ears. However, even as late as 1828, Thorburn's Seed Catalog carried only one variety of sweet corn.

As early as 1716 the Reverend Cotton Mather, who is remembered today mostly for his role in the Salem witch trials, observed how the wind could transfer the properties of red and blue corn from the windward part of a field to stands of yellow corn downwind by carrying the pollen to them. But of course nothing was known then about genetics. Flint corn, which was to spread over early 19th century New England, was not introduced until 1818 by Samuel Dutton of Cavendish, Vermont.

Broom corn, a special variety that many farms grew as their only cash crop, was sold for use in manufacturing the flat brooms that replaced the bundles of saplings used before. Its seed was also excellent feed for chickens.

One of the reasons that corn remained a favorite New England crop, as food for man and beast, was that in normal times it grew best in this northern climate. Yields per acre in New England were several times that in southern states such as

Corn, primarily used as fodder for livestock, was a staple crop in New England. The first white settlers learned to grow corn from the Indians, who used it in making bread and gruel. The short growing season in northern New England, however, made corn especially vulnerable to climatic variations, such as the snows and killing frosts in the summer of 1816.

South Carolina. Cold winters prevent the winter pupation of the corn-ear worm, discourage bacterial wilt and, when corn is cut for fodder, the colder weather delays the conversion of sugars to starches.

But the cold climate had its disadvantages. Along the Canadian border the corn growing season is normally only 120 days —compared to 240 along the Gulf of Mexico. The optimum growing season is between 150 and 180 days, and this is what defines the long narrow corn belt of today: Ohio, Illinois, Iowa and Kansas. Growing as it does in this narrow climatic zone, corn is susceptible to climatic fluctuation—as the cold summer of 1816 amply demonstrated.

The history of U.S. agriculture revolves around corn. It is still the largest crop in volume the country produces—approximately 4 billion bushels—far exceeding the crops of wheat, oats, rice and barley. Precise figures are not available for the total corn crop in the United States before the census of 1839, but in that year the total corn harvest stood at 378 million bushels, as compared to 85 million bushels of wheat, 123 million of oats, 4 million of barley, 18 million of rye and 9 million bushels of buckwheat.

Most of the corn was not consumed by man, but was used for sustaining cattle. Even in 1816 raising cattle for the urban market was an important source of farm income. In 1806, for example, 15,000 cattle were driven along the twisting winding dirt tracks, which in those days were called highways, from Vermont to Boston. However complacent many reports of the ill effects of the summer of 1816 on New England agriculture may seem, the loss of the corn crop was a bonafide disaster, particularly in the small isolated scrabble farms where self sufficiency and survival was the result of a delicate balance between man and the plants and animals he raised.

When killing frosts and snows upset that hard-won balance the summer of 1816, the hand-to-mouth north country farmers would have found it hard to understand, no doubt, that the cause was a volcano a half-world away.

CHAPTER 6

THE HARVEST OF 1816

Alarms and assessments

WITH FROST OCCURRING every month of the summer in New England and from the widespread reports of failure of the corn crop, the modern reader might expect that the inland farmers were in a state of panic. Of course there was no television to tell them that they ought to be frightened at the prospects. But they would themselves make up for this deficiency when they came to tell their grandchildren about the cold summer "1800 and froze to death."

By contrast, it is interesting to turn to farmer Harwood's diary once again, and see what he has to say about the fall of 1816 on his Bennington, Vermont, farm. From what he tells us, it would appear that his crops were fairly copious; he found time for many non-agricultural activities, for reading Roman history and for tinkering with his stove. Bennington was favored with a sheltered location and Harwood himself was a superior farmer. Certainly he manured his field, ploughed and harrowed with an eye to the weather and took advantage of the many dry days in September.

It is true that he wrote to Asa Harwood of Sempronius, New York, that the reason he was having difficulty in selling some furniture left with him was that money was extremely scarce because of the "badness of winter wheat crops, hay, etc."

As winter drew in Harwood made no mention of special hardship or deprivation. Visitors came and went at his farm;

71

from time to time he played his flute at a musical evening and always he found time to read. On Christmas Day, 1816, he butchered two hogs with a total yield of 470 pounds of pork and on New Year's Eve, "we threshed 3½ bushels of spring wheat which was all that remained in the barn."

It is difficult to piece together a coherent picture of the impact of the corn-crop failure from ephemeral accounts in the newspapers. Literate farmers were probably not representative of the mass of their fellow farmers. The climate in Bennington is not as harsh as it is farther north where pioneering, subsistence farmers might have suffered much more distress without leaving written records—with rare exceptions such as Reuben Whitten's epitaph (See page 105).

Throughout the fall of 1816, newspapers treated the weather and crop anomalies in mixed fashion. Although in many places corn and potatoes had been cut off and hay was very short, the American crops of English grain were very good. But wheat in Canada, still too green to resist the sharp frosts of September, also perished. The cultivators who had hoped to secure a large surplus quantity for the market did not even reap a sufficiency for their families.

The Halifax *Weekly Chronicle* noted that "great distress prevails in many parishes throughout (Quebec) Province from a scarcity of food. Bread and milk is the common food of the poorer classes at this season of the year; but many of them have no bread; they support a miserable existence by boiling wild herbs of different sorts which they eat with their milk; happy those who have even got milk and have not yet sacrificed this resource to previous pressing wants!"

Later, in December, the *Weekly Chronicle* reports: "It has been given us from the most authentic sources, that several parishes in the interior parts of (Quebec) are already so far in want of provisions, as to create the most serious alarms among the inhabitants. Among those mentioned as being in need of almost immediate assistance, we find a part of the Bay S. Paul, les Eboulements, St. Andre, Caconab and Rimbuskie."

At St. Johns, Newfoundland, 800 would-be immigrants were unceremoniously sent back to Europe because of a shortage of food in the town.

The November 16, 1816, *N. Y. Museum* carried a recipe for

potato bread using potatoes boiled and mixed with rye meal or flour, "and you may make as good bread as can be made from the best rye and Indian meal."

By February, 1817, times were hard enough in New York for the citizens to call a general meeting at City Hotel, where 10 committees were appointed to solicit subscription for relief of the poor. They collected $757 on the first day. Soup houses were opened to feed the starving. A new soup house at Franklin St. opened on February 14 and by March 15 had distributed 103,312 rations of soup and 1,000 pounds of navy bread (a pint and one half to a ration) "to the suffering poor." The poor were said to amount to one-seventh of the population. Volunteers manned the soup kitchens day and night, cooking around the clock in order to keep up with the demand.

Not all blamed the food shortcomings on the abysmal weather. The New York *Evening Post* of March 12, for example, blamed the problems on the city's 1,800 licensed tippling houses. Given the inadequacy of newspaper reports of the time, the variety of individual experience and the lack of government reports, how is it possible to determine what really happened?

Fortunately, three other sources of information on the social impact of the cold year are available: (1) a special enquiry instituted by the Philadelphia Society for the Promotion of Agriculture, (2) the record of wholesale prices for agricultural products, and (3) studies of patterns of emigration from the afflicted states to new lands opening in the West.

The Philadelphia Society

In response to widespread interest in the cold summer, the Philadelphia Society for the Promotion of Agriculture resolved on October 30 that the curators "collect facts relating to Agriculture and Horticulture, and of all circumstances connected therewith, which have occurred through the extraordinary season of 1816; and particularly the effects of Frost on vegetation, so far as it shall be in their power to acquire a knowledge of them . . . The result of such inquiries to be drawn into the form of a report."

Copies of this resolution were circulated widely by the secretary of the society, Roberts Vaux, a Philadelphia Quaker and

At the Brighton cattle fair held in October, 1816, there were few exhibits. An observer at the fair noted three reasons why the cattle were fewer and thinner than usual. "The unexpected season which almost threatened famine, the novelty of the exhibition . . . and a general but unhappy opinion that the great Springfield Oxen would be sent and would, of course, distance all competition."

prison reformer. Soon various replies to the resolution began to arrive.

One of the first was from Dr. Samuel Latham Mitchill, 53-year-old physician and Professor of Natural History at Columbia College of New York. In February, 1817, Mitchill was helping found the Lyceum of Natural History of the City of New York, destined to become the New York Academy of Sciences. Mitchill was its first president and, as author of a book about marine animals, presumably was responsible for the inclusion of boiled tongue of sea elephant (*Phoca elephantina*) on the menu of one of the Lyceum's early dinner meetings.

Mitchill summarized the situation:

> "There will not be half a crop of maize on Long Island, and in the southern district of this state. Further northward

there will be less. The buckwheat is so scanty, that a few days ago I paid four dollars for a half bushel of the meal, for the use of my family.

"The season, though unusually cool, was nevertheless warm enough to ripen strawberries, raspberries, currents, cherries, gooseberries, pears, plums, and apples. They were generally very fine. The ox-heart cherries, in particular, were unusually large and abundant. This autumn apples are fairer, cheaper, and more plentiful, than they have been for many years. Peaches were poor, owing to the distemper of the trees of several years standing.

"It is certain that the fruit has been damaged less by insects than is usual. An entomologist complained to me, a few weeks ago, that it has been a most unfortunate season for the collection of insects. That kind of game, he said, was so rare, that he had added but little to his museum.

"There have been at New York fewer fleas and mosquitoes than ordinary."

Another reply came from General David Humphreys, President of the Connecticut Society of Agriculture. At the age of 64, Humphreys was a hero of the revolution and had carved chicken at Mount Vernon. A poet, soldier and statesman, he had served Washington in delicate diplomatic missions and as a secret agent in Portugal and Spain.

Humphreys wrote: "The principal injury done by early and late frosts fell on our most important crop, Indian corn." Corn had ripened so poorly, the general noted, not more than a quarter of what was sown was usable for meal. The rest, unripe, mouldy and soft, was fed to hogs and cattle, but did not seem to fatten the livestock. Grasses for animal fodder were diminished by about 50 percent.

Turnips, carrots and other root crops "have generally been more flourishing and productive than in ordinary seasons." This may have been because, despite the drought, more dew fell than usual.

While the grub worm "was never before so frequent and mischievous," the canker worms and caterpillars, which "proved so pernicious to orchards" in the past, were destroyed.

The consensus was that although there was little corn, certain fruits and vegetables had actually benefited by drought and cold.

The national enthusiasm for labor-saving machinery was still in its infancy. But the 1816 weather and the resulting shortage of animal feed provided stimulus for development of one farm machine: The Dover, New Hampshire, Sun *carried a long article on using a "chaff cutter" to chop up corn stalks for fodder. "A powerful straw cutter is essential. Additional labor is, no doubt, required; but the important savings of provender quadruply repays it."*

Unfortunately, the Philadelphia Society's report contained no responses from the northern states of Vermont, New Hampshire or from Canada.

A Quaker correspondent who identified himself as a "humble ploughman," petitioned the Philadelphia Society on the behalf of wild birds, amongst whom the cold summer had made "dreadful havoc." He urged that sportsmen abstain from shooting blackbirds, robins, jays, catbirds and others so that their numbers could increase again to the point where they would resume their usual role in keeping down the insect pests.

The celebrated English radical and polemicist, Thomas Cobbett, was an innovative experimental farmer. In 1817, he came to America for the second period of self-exile—the first had ended when he was fined for accusing Dr. Benjamin Rush of bleeding George Washington to death. This time he hired a farm in North Hempstead, Long Island, confounding his neighbors by transplanting corn, cultivating rutabaga and burning earth.

The cold summer of the previous year could scarcely have loomed large in local conversation, otherwise Cobbett would not have entered in his diary, on June 1, 1817: "Fine warm day; but saw a man, in the evening, covering something in the garden. It was kidney-beans, and he feared a frost." Where are the snows of yesteryear?

An early conservationist

At this point it might be interesting to add the observations made by one particularly acute observer of the passing scene: Noah Webster. He had an obsessive interest in facts of all kinds.

Compiling a dictionary of the English language was just one of Noah Webster's interests. He also liked to count houses in the cities he visited and keep data on weather.

One might have encountered his slender form walking with hands clasped behind his back, counting houses during a visit to a strange city: He counted 3,500 in New York City during a sojourn there. Besides his determination to list all the words of the English language, Webster had an interest in vital statistics and their relations to public health, epidemiology and the insurance business—and an interest in weather and climate records and their relation to agricultural production and famine.

With the cold summer of 1816 behind him—and uncannily echoing today's concerns—Webster wrote in *Spooner's Vermont Journal* on May 5, 1817, an article on the energy crisis that he anticipated would soon strike New England. He pointed out that using wood for fuel would soon outrun its annual replacement growth and that unless Americans were to "delve into the bowels of the earth" for substitutes, conservation measures would have to be undertaken. He recommended using more efficient iron stoves, allowing green wood to dry thoroughly before use, narrowing flues in chimneys and building new houses with an eye to reducing heat losses and using scrap wood and chips more extensively.

CHAPTER 7

THE MACKEREL YEAR

Inflation of the price of food

SHORTAGES DUE TO crop failures in New England, Canada and parts of Europe caused the price of wheat, grains and flour to rise sharply.

By June 19, 1817, the *New York Post* protested, "the high prices which meats, vegetables, butter, milk, and in short everything in our market continue to bear can be viewed in no other light than the greatest of impositions on our citizens . . . and call . . . for some general measure of redress."

The following table shows annual mean prices for wheat in New York:

LIST OF WHEAT PRICES TAKEN AT ALBANY ON 1 JANUARY

Year	Shillings* per Bushel	Year	Shillings per Bushel
1808	9	1817	18
1809	8	1818	15
1810	12	1819	14
1811	14	1820	8
1812	15	1821	6
1813	18	1822	9
1814	15	1823	10
1815	13	1824	10
1816	14	1825	8

(* Eight shillings equal $1.00)

An early 19th century drawing of Albany, New York, by J. H. Milbert. Wheat prices recorded in Albany in 1817 following the terrible growing season of 1816 were as high as those brought on by the War of 1812.

We see in this table that the price of wheat rose to a maximum at the beginning of the War of 1812. It hit 18 shillings a bushel January 1, 1813, and then fell, rising again in 1817 to a second peak, which presumably was due to the crop failures. The effect seems to have been equivalent to that of a new war.

A reference book on wholesale prices in Philadelphia quotes Grotjan's Philadelphia Public Sale Report (November 18, 1816) as follows:

> "This article (flour), as will be observed by the general prices current, has experienced a great and sudden rise; owing to intelligence received from Europe that the crops of grain have failed in several countries, it is to be expected that large demands will be on our surplus. As our crops of wheat and rye have been considerable, this surplus might perhaps have been sufficient to answer the expected foreign

Prices of wheat, corn, pork and beef in New York in 1816. They reflect the shortage of agricultural products brought on by crop failures and the glut of animal products, due to the wholesale slaughtering of livestock. Farmers were faced with either allowing their animals to starve to death or going bankrupt buying animal fodder. See page 84.

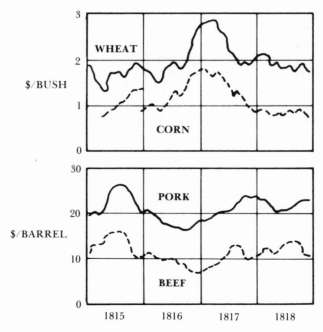

demand, without unreasonably enhancing the current prices of former weeks, if not a great portion thereof was necessary to compensate for the lamentable deficiency of the crops of Indian Corn which, in consequence of the remarkably cool season experienced during the summer months, have been greatly deficient in the states of New Jersey and Pennsylvania, and perhaps more so in the states of Maryland and Virginia; few farmers having been able to raise sufficient for their own use."

A 19th century weather historian, Sidney Perley, gives us the following assessment of the effects of the cold summer on prices:

"There was great destitution among the people the next winter and spring. The farmers in some instances were reduced to the last extremity, and many cattle died. The poorer men could not buy corn at the exorbitant prices for which it was sold.

"In the autumn, stock was sold at extremely low prices on account of lack of hay and corn, a pair of four-year-old cattle being bought for thirty-nine dollars in Chester, New Hampshire.

"Some favored spots in the northern New England states produced a little corn for seed, which commanded a great price the following spring. Abraham Sargent, Jr., had removed from Randolph, Vermont, to his father's farm in Chester, New Hampshire, and brought with him a very early kind of corn. He raised a crop of tolerably sound corn which he sold for seed the following spring at four dollars per bushel, and the farmers esteemed it a great favor to obtain it at that price even.

"The next spring hay was sold in New Hampshire in a few instances as high as one hundred and eighty dollars per ton, its general price, however, being thirty dollars. The market price of corn was two dollars per bushel; wheat, two dollars and a half; rye, two dollars; oats, ninety-two cents; beans, three dollars; butter, twenty-five cents per pound; and cheese, fifteen cents. In Maine, potatoes were seventy-five cents per bushel, the price in the spring of 1816 having been forty cents, which was the usual price. Pumpkin seeds were sold in Massachusetts for one dollar per hundred, and other seeds proportionately. Fine crops of Indian corn were raised for many successive years following this cold summer of 1816.

"Ever since that cold year, old people have continued to tell about its unfruitfulness, and some of their stories were exaggerated as stories will become by repetition. For example, Jacob Carr of Weare, New Hampshire, used to boast of the large crop of potatoes that he raised that year, and said that he did not get less than five hundred bushels to the acre, and that he never allowed one to be picked up that was smaller than a tea-kettle."

One of the most complete studies of the effects of the cold summer on the price of agricultural products is the compilation of monthly wheat, corn, beef and pork prices for the years 1814 to 1818, made by Joseph B. Hoyt from an examination of

The price of wheat did not again reach the 1816 peak until the Russian wheat shortage of 1972 and the Arab oil embargo of 1973. This is even more remarkable considering that, despite inflation, the cost of agricultural products had been steadily decreasing due to modern machinery and scientific farming techniques.

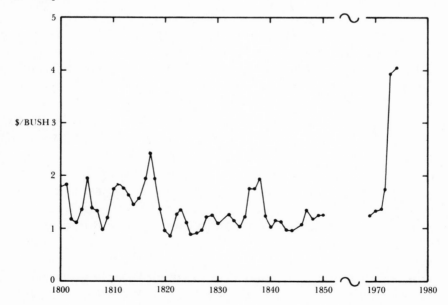

the price reports found in New York newspapers for every week. Hoyt used the *New York Spectator* for 1814, 1815 and 1818, the *New York Herald* for 1816 and 1817, in order to get complete weekly coverage for the five-year period. Even then there are a few gaps and discontinuities. These prices are depicted in the graph on page 81, which shows prices of wheat and corn in dollars per bushel, and of pork and beef in dollars per barrel. The most striking feature of the wheat and corn prices is the great rise in prices starting about October, 1816, when it became apparent that the corn harvest would be poor, and its drop in the summer of 1817 when the new crop was obviously going to be plentiful.

On the other hand, the prices for pork and beef began to fall noticeably by the middle of 1816 when farmers began to sell their stock in anticipation of a poor hay yield—and began to realize that they could not hold their herds over the fall and winter due to prohibitive prices for fodder. The normal peak for beef prices was in August, just before farmers ordinarily drove their livestock to market—these peaks can be seen for the normal years 1815, 1817 and 1818. But the poor prospects for hay, which were apparent early in 1816, completely wiped out the normal August maximum. Despite the upset, the price structure had returned to normal by the fall of 1817.

If we look at average yearly prices of wheat over many years we see how very special the year 1816 was economically. The prices of wheat in dollars per bushel are shown in the graph on page 83, plotted from figures tabulated by the U.S. Bureau of the Census for the 50-year period ending in 1850. It is remarkable how steady this price has been over the years. Somehow the great increase in productivity of agriculture managed to keep even with overall inflation during this period. Between 1800 and 1850, 1817 is clearly the highest price, reflecting the high winter and spring prices of early 1817.

For comparison look at modern-day prices. From 1969 to 1972, wheat prices are remarkably similar to those of more than a century earlier. The early 19th-century wheat prices would only be finally topped after the great Russian wheat crop failure of 1972 and the Arab oil embargo and oil-price increase of late 1973! At first look, the 1972 increase appears to be very much greater than the price jump that occurred in 1816–1817 —until one reflects that the dollar was worth perhaps 20 to 40 times more back then. Food was so costly in 1816 that any tiny increase would be severely felt in the budget of a non-agricultural worker.

In Monticello, Thomas Jefferson, who was not a very successful farmer despite all his other great attributes, was so affected by the poor corn harvest that he was reduced to applying to his agent, Patrick Gibson, for a $1,000 bank loan. Poor farmers in the northern counties of Vermont and New Hampshire were unable to feed their pigs. By early spring they were catching fish with seines across the northern rivers or importing mackerel from the seaports. Thus 1816 by some was remembered as the "Mackerel Year."

Government response

During the summer and fall of 1816 the federal government in Washington took no official notice of the growing agricultural crisis in New England and elsewhere. Congress was not in session. Government buildings were being restored after their recent burning by the British. The destruction had been carefully limited to public buildings according to instructions of the British commander, and the Patent Office had been spared on the technical grounds that the patent models it contained were private property. The president's house had been gutted and, during the restoration the gleaming white paint that covered the scars triggered the first use of the name "White House."

President Madison had little time to spare for worry over crops in the northern states. There was the demilitarization of the Great Lakes to work out with the new British minister, Sir Charles Bagot. The French minister, Baron Jean Guillaume Hyde de Neuville, needed to be soothed after having become enraged at the toast of the postmaster of Baltimore at a Fourth of July celebration in Baltimore, which referred to Louis XVIII as that "imbecile tyrant." And, to cap it off, Nicholas Kosloff, the Russian consul general, was arrested in Philadelphia for the rape of a 12-year-old serving girl. Questions of diplomatic im-

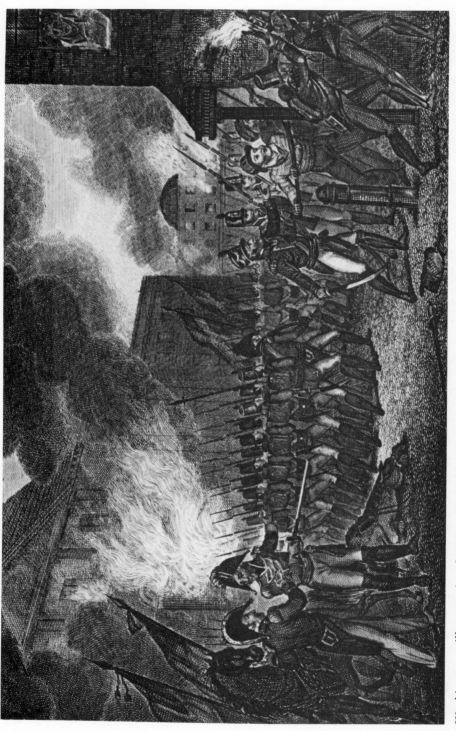

Washington, still recovering from its capture four years earlier by the British, did not pay much attention to the plight of New York and New England farmers suffering the ill effects of the abysmal summer weather of 1816.

munity were raised by the Czar's government, and the American chargé in St. Petersburg was made subject to reprisals and was banned from court. It would not be until the opening of the next Congress, late in the year, that the cold season in New England would be referred to the federal government.

In New Brunswick, Canada, the governor temporarily suspended import duties on grains and prohibited their export or distilling them into spirits.

Among the acts of Nova Scotia's spring general assembly of 1817 was one to prohibit the exportation of wheat, rye, barley, Indian corn, oats or potatoes from the province for four months.

That winter similar legislation was proposed in the United States Congress. Following James Madison's address in December, 1816, Mr. Stephenson Archer "presented a petition of sundry inhabitants of Harford County, in the state of Maryland, praying that a law be passed prohibiting, for a limited time, the exportation of breadstuffs, and the distilling of spirituous liquors from grain, in consequence of the short crops of those articles made at the last harvest, and the high prices they now bear." This petition was referred to the Committee on Commerce and Manufactures. Nothing further was heard about it, perhaps because it was Friday the thirteenth or maybe because it was an obvious ploy on the part of temperance interests.

Further, on January 29, 1817, a petition of the Berkshire Association for the Promotion of Agriculture and Manufacture in Massachusetts was presented to Congress "praying that the aid of the National Government may be extended to the promotion of agriculture . . . either by the establishment of a national board, or by such other measures as in the wisdom of Congress may seem meet and proper." This was referred to a select committee, and was later read twice and referred to the Committee of the Whole. On February 20, 1817, Mr. Charles Goldsborough of Maryland resolved "that the Committee of Commerce and Manufacture be instructed to inquire whether any, and if any, what measures may be necessary to be adopted in consequence of the great failure of the corn crop in the past year."

One of the side effects of the cold summer of 1816 was it hastened the construction of better transportation facilities to improve trade in interior regions. The Erie Canal was started in 1817. This drawing shows the excavation techniques used in the construction of the canal.

Life in upper New York State quickly revolved around the Erie Canal, as shown in this 1830s Tombleson engraving of Lockport in N. E. Willis' American Scenery.

The Erie Canal and others begun after the disastrous summer of 1812 did improve inland transportation. But as early as 1816, when canals were being discussed seriously, more visionary planners could see they would prove too cumbersome and slow to meet America's long-range transportation needs.

Congress did not seem to believe there was any very great emergency, and even refused to offer any aid to the construction of the Erie Canal. Evidently New Yorkers felt a more urgent need for developing inland transportation, because the canal was authorized by the New York Legislature in April of 1817. Work began on the first 15 miles between Utica and Rome on July 4, 1817. This section was opened two years later. It was October 26, 1825, before the entire canal (363 miles long, 40 feet wide and 4 feet deep) was opened with appropriate official celebration. The canal was a great success, but was eventually overtaken by railways, as John Stevens of Hoboken had predicted in 1812. Stevens was the father of the first U.S. Patent Law, pioneer in application of steam power and inventor of the screw propellor. He could see, even before the first railroads had been built, that canal travel would prove too slow and cumbersome "to satisfy the demands of humanity."

It was only many years later that Congress made effective moves toward supporting agriculture. Thus farmers had to wait until 1862 when the Morrill Act established agricultural and mining land grant colleges (Cornell and Michigan State University being the first) and led to the widespread education necessary to encourage practical use of scientific knowledge on the farm. If one were to gauge the gravity of the crop shortfall in 1816 by the degree of government response to it, one would find it hard to justify the prominence which the year 1816 acquired years later in local legend.

CHAPTER 8

FLEEING THE SCRABBLE FARMS

Westward migration

By 1816 THE most fertile land of northern New England was fully settled. As farm families were large, there were many more sons than could possible inherit the land. Considering the relative insignificance of industrial enterprise, there were few boys who could stay in northern New England and find employment. The only possibility of increasing prosperity was to emigrate to new lands. Active, industrious and enterprising young men were bound to move to those new parts of the continent that offered open territory. This was the constant and underlying reason for emigration to the west in the first half of the 19th century.

The summer of 1816 marked the point at which many New England farmers who had weighed the advantages of going west made up their minds to do so. For several years, letters from earlier emigrants enticed the youth of Vermont, New Hampshire and Maine to the fertile flat farmlands between the Ohio and the Mississippi rivers.

Now more decided to make the break. Some moved in groups of new religious sects. Others moved from the farms of the northern states to the mills and industries that were springing up in southern New England. Still others were to be counted amongst the "green Vermont and New Hampshire

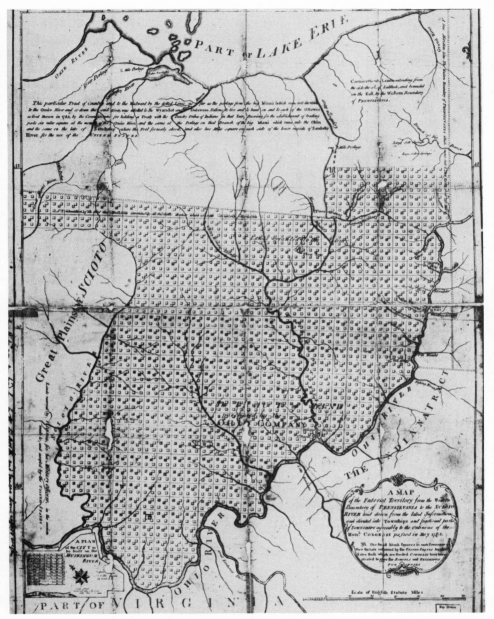

This is Cutler's Map of Ohio, published by Manasseh Cutler, a major promoter of the Ohio Company of Associates, illustrating a huge eastern Ohio subdivision planned in the wake of the Federal Land Ordinance of 1785. Projects such as these helped lure tens of thousands from the east following the intemperate summer of 1816.

country boys" whom observers like Herman Melville were later to see arriving at the whaling-ports along the coast, anxious to try their hand at the great whale fisheries of the South Seas.

Vermont as a case study

In order to assess quantitatively the effect of the 1816 weather on the population, let us turn to the particular case of Vermont.

The gradual approach of population toward equilibrium since the time of Vermont's settlement is illustrated by the accompanying graph. The population figures are taken from the Federal Census between 1790 to 1860. It has the familiar exponential form of a population with fixed economy and steady renewable resources, leveling off at about 350,000. In the decade 1810 to 1820 there is a marked decrease in the rate of increase of the population. It looks like a rather minor dent in the curve, but it actually represents a loss of 10,000 to 15,000 people, who otherwise would have stayed in Vermont.

Population of Vermont 1790–1860

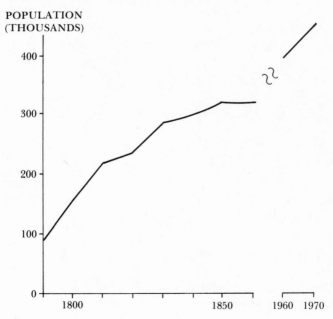

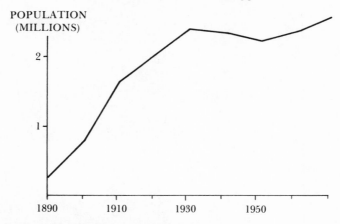

POPULATION
(MILLIONS)

2 –

1 –

1890   1910   1930   1950

Population of Oklahoma 1890–1970

It set the growth of Vermont back about seven years. Those lost may well have been from amongst its most enterprising citizens.

In order to illustrate the magnitude of this population dislocation, look at the upheaval in Oklahoma during the Great Depression, when drought turned that state into a dust bowl. Previous to 1930 the population of Oklahoma was climbing steadily, doubling every 15 years. When the sharecroppers fled their parched fields the population of the state actually dropped. The state lost 14 percent of the population that otherwise it might have anticipated. This contrasts with about 8 percent for Vermont between 1810 and 1820. The drought in Oklahoma lasted longer than the one cold summer of 1816. The Dust Bowl effects were cumulative, and it was not until 1960 that the population of the state rose again to what it had been in 1930.

Emigration was uneven within Vermont and Maine. Some townships were hit harder than others. Granby, Vermont, for example, was a considerable town in 1800 and had grown rapidly by 1810. The cold season of 1816 had such a depressing influence that only three families were left in 1817. It wasn't reestablished until 1821.

Vermont historian L. D. Stillwell examined diaries of people migrating to the west from Vermont and showed that in 1816–

In many ways, the emigration from the Dust Bowl during the 1930s resembled the emigration west from New England following the privations of 1816.

1817 the number of migrants was almost double that of the remainder of the decade (See page 98 for graph).

Because Stillwell's figures are for individual years, instead of for decades, as in the Census, we can establish the annual fluctuation of emigration. The graph is based on a so-called relative rate of emigration between 1810 and 1820. It has a notable peak in 1816. The decision to move westward was evidently a quick one, as witnessed by the Zanesville, Ohio, *Messenger* of October 31, 1816:

> "It is remarked . . . that the number of emigrants from the eastward the present season, far exceeds what has ever before been heretofore witnessed. It is impossible to calculate to a certainty the number of persons who have passed through this place within a few weeks past; but it must amount to some thousands besides those who have descended the Ohio, or taken other routes by lands. On some days, from forty to fifty wagons have passed the Muskingum at this place. The emigrants are from almost every state north and east of the Potomack, seeking a new home in the

... territories of the west; travelling in various modes—
some on foot, some on horses, and others in different kinds
of vehicles, from the ponderous Pennsylvania wagon, to the
light New England pleasure carriage."

This stream of humanity added 42,000 settlers to Indiana in
1816.

Those who left Maine

In his *History of Maine Agriculture,* Clarence Day devotes a
whole chapter to the flight of Maine farmers westward.

> "All through the cold year and the next, emigration con-
> tinued active. Companies formed to buy land in Ohio and
> Indiana advertised for settlers. Farmers sold their farms and

*Unlike the covered wagons of the early 19th century, refugees of the Dust Bowl were able
to make 100 miles a day or more, as these two families from Missouri looking for work in
the California pea fields.*

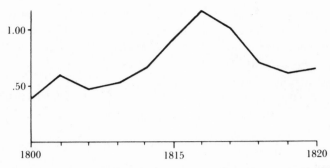

The relative rate of emigration from Vermont, based on a study of diaries, shown on pages 95–96.

long journey overland to the Western Land of Promise. Some of the poorer people even trudged the weary way on foot. One such family of eight, bound for Indiana, walked all the way from Maine to Easton, Pennsylvania, where they arrived in dead of winter dragging their few possessions and the smallest of the children in a hand cart. On a single day a train of 16 wagons with 120 men, women, and children from Durham, Maine, passed through Haverhill. Their minister accompanied them, and they were going to Indiana to buy a township. One day the editor of the *Genessee Farmer* met a train of 20 wagons and 116 persons on their way to Indiana, all from Maine."

The restless population's emigration to the west led to the first great era of land sales and land speculation. In New York and Ohio there were huge unsettled areas that speculators had managed to obtain through various forms of favoritism and opportunism. Many landed proprietors moved to the New York frontier. Amongst the Virginia planters who moved to the region surrounding Seneca Lake, was the father of James Fenimore Cooper, who settled in what was to be called Cooperstown.

Many large properties were rented out to tenant farmers, a practice that was alien to New England. Land that could be purchased from the Holland Land Co. for $1.25 to $4 an acre could be partly paid for by selling the ashes from the felled trees as potash, which sold for as much as $90 a ton.

There were many causes for the land boom. The summer of

Life in the Ohio frontier was hard and desolate—even compared with New England's remote scrabble farms—as shown in this circa 1810 illustration from Pioneer History of the Holland Purchase *by O. Turner.*

1816 was only one of them. A bounty of 160 acres was offered to every veteran of the War of 1812. Those who sold their bounties to real estate operators encouraged speculation. The scramble for land was accelerated by loose credit extended by the government. Land speculation was further encouraged by a favorable foreign trade balance for American agricultural products and the prospect of public expenditures on roads and canals promised to make the land accessible. By 1817 land sales had doubled in two years and were to double again by 1819.

The loss of population became so alarming that letters began to appear in newspapers arguing the merits of Maine, and the defects of the western country. Every conceivable argument in favor of Maine was offered; the absence of malaria, the nearness to Europe—even the good sledding in winter.

So great was the concern in Maine that its population would be lured west, that the depressing accounts of life in the west were given more space by the *Argus* than a lurid account of a Kentucky church struck by lightning while its congregation was worshipping, a revolt of Jews in Cairo, or even descriptions of several sightings of a sea serpent off Cape Ann, lying on the surface of the water, coiled up in serpentine form, reposing in conscious security, after a hearty breakfast of fresh herring.

William Brown, of Bath, was one of the many Maine farmers who had a sad tale to tell upon his return from the west in the summer of 1817. He had put the deed of his farm in escrow with a land agent in April 1817. It entitled him to $1,000 credit toward buying a new farm in western Pennsylvania. Travelling to Philadelphia by sea, and then proceeding inland, he found the welcome extended to Yankees decidedly cool: the locals were too accustomed to being victimized by sharp Yankees.

He also found life distressingly primitive compared to Maine —bear skins rather than feather beds, the rough and tumble of frontier biting-and-gouging fights and the lack of bread, butter and cheese. A 150-acre farm that was offered to him at $6 an acre had only four acres cleared, and an abandoned log house. And when he discovered that the corn crop of 1816 had also been killed in western Pennsylvania, he decided to come home.

CHAPTER 9

TALL TALES AND EXAGGERATIONS

Legend and folklore

LEGENDS ARE PART of collective folk history. They preserve accounts—distorted as they may be—of singular events that at the time seemed important but that go unnoticed in standard histories. Legends are like the sad, blurred daguerrotypes of long-gone ancestors in family albums; they provide little glimpses into the past. Too incomplete and inexact to document the past, legends nonetheless lure us all into trying to do so. They catch our interest—that is what they are intended to do.

The man who "froze to death"

With the passage of time the truth is embroidered, or at any rate selected or altered by those telling a story for maximum effect. Despite talk of how cold the summer was, there appears to have been only one cold-related death. James Winchester reminisced in a newspaper article at the age of 90 about the only known fatality of the year without a summer.

> "I was 14 years old then, and lived in Vermont, where I have always lived, and where that memorable season was at its very worst.

"An uncle of mine had some sheep in a back pasture lot. To get to that lot he had to go through a piece of woods for nearly a mile. The weather had been very cold all through June. The big storm of the 17th began along about noon, and my uncle started after dinner to go to the sheep-pasture to fix up a shelter of some kind for the sheep. No one had any idea, cold and eccentric as the season had been up to that time, that we could have a fall of snow that would amount to anything at that time of the year.

"I was at my uncle's when he left home to go to the sheep lot, and as he went out the door, he said to his wife in a jocular way: 'If I'm not back in an hour, call the neighbors and start them after me. June is a bad month to get buried in the snow, especially when it gets so near the month of July.'

"The snow increased in fury, and by night it drifted so that the roads were almost impassable, but even then and when it grew dark, none of us felt uneasy about uncle. The weather had become bitter cold. When night set in in earnest, and there was no sign of my uncle's return, his wife sent me and my cousin, who was two years younger than I, to alarm the neighbors and tell them that we believed uncle had been lost in the snow and perished.

"We had a hard time getting to the nearest neighbor's, less than a mile away, and there gave the alarm, but could go no further. The neighbors summoned others, and in spite of the severity of the night, they searched the woods until morning, but no sign of the missing man could be found. The search was taken up by others the following day, and all the next night, without any trace of his being discovered, except that he reached the pasture and built a shelter of boughs in one corner of the lot, under which the sheep were huddled. On the forenoon of the third day the searchers found my uncle buried in the snow a mile from the pasture, in an almost opposite direction from the home. He was frozen stiff . . ."

When faced with this account we must allow for the extreme age of the narrator and the long time that had transpired between the event and the telling. Also, June 17 seems too late a date. Elsewhere the snow was earlier in June. In the absence of an autopsy, might not the uncle simply have had a heart attack while roaming the blustery fields?

The only authentic case of bodily injury seems to have been that of 88-year-old Joseph Walker of Peacham, Vermont, who, lost in the woods on the night of June 7, so froze his toes that they had to be amputated.

More obvious misstatements of fact can be found easily, for example the following oversimplification in the September 6, 1924, issue of *The Literary Digest:* "In 1816 there were no crops raised north of the Ohio and Potomac and but scanty returns much further south. Frost, snow, and ice appeared in every month of spring, summer, and fall."

Samuel Hopkins Adams's tall tale

A more amiably embroidered example is an article by Samuel Hopkins Adams in *The New Yorker* of May 19, 1956, entitled "Grandfather and the Cold Year," which purports to be a verbatim account of a conversation of the author as a child with his grandfather in the sitting room of his cottage on South Union Street, Rochester, New York, on a chilly day in May, 1883. Grandfather recounted to his grandchildren how in 1816 he was a sophomore at Hamilton College (the college records indicate that he did not enter until October of 1817) and how they lived through the following winter on mackerel which, being sold undersalted by the rascally fishmongers of New York City, poisoned enough people to persuade Governor DeWitt Clinton to appoint "mackerel inspectors."

A search of the New York City Council minutes of 1816 and 1817 for references to New York City fishmongers does show concern over the Fly Market's nearness to an open sewer and plans for opening a new (the Fulton) fish market—but no mention is made of special concern about fish inspection. Perhaps the mackerel inspector operated upstate along the Hudson.

Adams recalls that his grandfather spoke of portents of the cold summer, such as immense flocks of passenger pigeons, which on their usual northward migration suddenly veered back southward in such quantity that they were knocked down in hundreds from their roosts and pickled in hard cider. He recalls hairy-bears (caterpillars) with fur so long they had difficulty crawling and of Obediah Dogbery, who surprised a polar bear while fishing at the Lower Falls of the Genesee River. Adams produces a journal in which is written: "May 26—Indians take a hand. Medicine men of the Senecas announce that the sun is mortally taken sick and can be cured only by a three day tribal pow-wow. Pow-wow held but no improvement in the sun's condition." And, "June 1—Dr. Hashalew, a well-known itinerant medical professor, announces that the seasons have exchanged places and advises all citizens to improve their woodpiles as for winter, and to lay in supplies of his Elixir Vitae, four shillings a bottle. Many do so."

The article continues to describe how grandfather, having been rusticated from college for pranks with the college bell aimed at prolonging the time they could stay in bed, cornered the asafetida market. He initiated a wolf-repellent insurance business to ward off the packs made so hungry by the unseasonable summer from neighboring farmers' sheep and hens.

The reference to wolves is not wholly facetious. In March of 1816, four Maine townships voted bounties on wolves up to $40. This was the last compliment paid by the towns to these animals, which subsequently entirely disappeared. But deforested and blackened mountain peaks, burnt over by farmers to drive away the wolves, brooded over the New England landscape for many decades more.

From county historians

It is difficult to know what to make of some of the more startling recollections quoted in county histories, such as the tale of how farmer James Gooding killed all his cattle and then hung himself—presumably in desperation over the gloomy weather. Then there is a tale by Anna Gilbert about how her father and other Dorset, Vermont, farmers slaughtered thousands of sheep at Pea Street Brook. Another is recounted by Bakerfield historian Elsie Well, who describes how Nathaniel Foster saved his cornfield by cutting and burning pine night and day. Another yarn that seems impossible to verify is about Shakers in the Berkshires who, influenced by a prophetess, began to hoard for a seven-year dearth on August 1.

Reuben Whitten's tombstone in an abandoned cemetery in Ashland, New Hampshire.

1816.	JULY hath 31 Days.

ASTRONOMICAL CALCULATIONS

Days.	☉ in	d.	m.	Days.	☉ in	d.	m.	Days.	☉ in	d.	m.
1	♋	9	26	13	♋	20	52	25	♌	2	20
3		11	20	15		22	47	27		4	14
5		13	15	17		24	41	29		6	9
7		15	9	19		26	36	31			4
9		17	3	21		28	30				
11		18	58	23	♌	0	25				

(O's place.)

☽ First Quarter 3d day, 4h. 44m. morning
● Full Moon, 9th day, 7h. 37m. morning.
☾ Last Quarter 17th day, 8h. 2m. morning
● New Moon 24th day, 6h. 25m. evening
☽ First Quarter 31st day, 9h. 41m. morning

M.D.	W.D.	☉ rises & sets		L.D. H. M.	D.dec. H. M.	☉ S.A.	☽ F. sea. H. M.	☽'s place	☽ rises & sets		☽ jou. H. M.	
1	Mond.	4	28	8 15	4 0	2 3	6 4	54	reins	morn.	5	39	
2	Tuefd.	4	29	8 15	2 0	4 4	7 5	43	reins	0 12	6	28	
3	Wedn.	4	29	8 15	2 0	4 4	8 6	32	reins	0 38	7	17	
4	Thurf.	4	29	8 15	2 0	4 4	9 7	22	fecrets	0 54	8	7	
5	Friday	4	30	8 15	0 0	6 4	10 8	13	fecrets	1 21	8	58	
6	Satur.	4	30	3 15	0 0	6 4	11 9	7	thighs	1 54	9	52	
7	SUN.	4	31	8 14	58 0	8 4	12 10	2	thighs	2 31	10	47	
8	Mond.	4	31	8 14	58 0	8 5	13 10	58	knees	3 16	11	43	
9	Tuefd.	4	32	8 14	56 0	10 5	● 11	53	knees	●rises.	morn.		
10	Wedn.	4	32	8 14	56 0	10 5	15 morn.		legs	8 48	0	38	
11	Thurf.	4	33	8 14	54 0	12 5	16 0	45	legs	9 21	1	30	
12	Friday	4	34	8 14	52 0	14 5	17 1	34	legs	9 50	2	19	
13	Satur.	4	34	8 14	52 0	14 5	18 2	20	feet	10 14	3	5	
14	SUN.	4	35	8 14	50 0	16 5	19 3	3	feet	10 36	3	48	
15	Mond.	4	35	8 14	50 0	16 6	20 3	44	head	10 58	4	29	
16	Tuefd.	4	36	8 14	48 0	18 6	21 4	24	head	11 16	5	9	
17	Wedn.	4	37	8 14	46 0	20 6	22 5	4	head	11 40	5	49	
18	Thurf.	4	38	8 14	44 0	22 6	23 5	45	neck	morn.	6	30	
19	Friday	4	39	8 14	42 0	24 6	24 6	29	neck	0 3	7	14	
20	Satur.	4	39	8 14	42 0	24 6	25 7	16	arms	0 32	8	1	
21	SUN.	4	40	8 14	40 0	26 6	26 8	7	arms	1 7	8	52	
22	Mond.	4	41	8 14	38 0	28 6	27 9	2	breaft	1 54	9	47	
23	Tuefd.	4	42	8 14	36 0	30 6	28 10	0	breaft	2 48	10	45	
24	Wedn.	4	43	8 14	34 0	32 6	● 11	1	breaft	☽ sets.	11	46	
25	Thurf.	4	44	8 14	32 0	34 6	1 eve.	1	heart	8 33	eve.	46	
26	Friday	4	45	8 14	30 0	36 6	2 1	0	heart	9 2	1	44	
27	Satur.	4	46	8 14	28 0	38 6	3 1	54	belly	9 33	2	39	
28	SUN.	4	47	8 14	26 0	40 6	4 2	46	belly	10 2	3	31	
29	Mond.	4	48	8 14	24 0	42 6	5 3	37	reins	10 30	4	22	
30	Tuefd.	4	49	8 14	22 0	44 6	6 4	26	reins	10 54	5	11	
31	Wedn.	4	50	8 14	20 0	46 6	7 5	17	fecrets	11 22	6	2	

Contrary to folk legend, the Old Farmer's Almanac *of 1816*

JULY, feventh Montn.   1816.

Soon as the morning trembles o'er the sky,
And, unperceiv'd, unfolds the spreading day,
Before the ripen'd field the reapers stand,
In fair array, and swell the lusty sheaves.

M. D.	W. D.	Courts, Aspects, Holidays, Weather, &c.	FARMER'S CALENDAR.
1	2	*Rather* ☽ perig.	Socrates said that "we should eat
2	3	C.P.&S.Bost. Visit.V.Mary.	and drink in order to live; instead
3	4	Mid. tides. *dull hazy*	of living, as many do, in order to
4	5	INDEPEN dec.1776. *weather.*	eat and drink." This is most ex-
5	6	*Now*	cellent advice, and I have but one
6	7	7's rise 1h. morn. ♂ ☽ ♅	word more to say, that is, drink
7	D	4th Sun. past Trin. *expect*	neither too much cold water, nor
8	2	*good hay weather*	too much hot rum.
9	3	S J.C.Plym. C P.&S.Bang.	
10	4	*for some*	Cut your grass in season. There
11	5	*days.*	is a size or glutinous matter upon
12	6	*Changes to*	your grass, of much consequence to
13	7	*dull.*	its sweetness and nourishing qual-
14	D	5th S. past Trin. Fr. Rev.	ity. The rains will deprive it of
15	2	☽ apogee. [com. 1789.	this excellence, unless you pay close
16	S	Very low tides. *Fine*	attention to the time of cutting and
17	4	*again.*	making your hay.
18	5	Mahomet died 634, aged 64.	
19	6	*Some*	Neglect not to hoe your cab-
20	7	Margaret. *showers*	bages. Hill your corn, when the
21	D	6th Sun. past Trin. *Rather*	weather will not permit haymak-
22	2	Magdalin. *catching*	ing. Gather your seeds as they are
23	3	7's rise midn. *hay*	ripened. Gather herbs while in
24	4	*weather.* ☐ ☉ ♃	bloom. This business is generally
25	5	St. James Dog days begin.	thought to be of little or no conse-
26	6	St Anne. Com.Bur.Univ Vt	quence ; but I tell you, friend, it is
27	7	Mid. tides. *A storm*	often the case that a bowl of herb-
28	D	7th Sun. past Trin. ☽ perig	drink, with a good nurse to attend
29	2	Y'd L. rises 3h morn. *is*	you a few hours, saves *you* from a
30	3	*not far*	fever, and your *purse* from a doc-
31	4	*distant.* ♂ ☉ ♀	tor's bill.

did not predict unseasonable frosts in the first week of July.

A story with the ring of truth in it is told by W. J. Bigelow in his *History of Stowe* (1934): "Joseph Fuller, Esq., at the Old Folks Festival held in Stowe, Vermont, on Sat., Sept. 20, 1874, told that in the years 1815 and 1816 not an ear of corn was raised in Stowe, and (that) there was snow on the mountains every month except July. 'The year when the crops were cut off my father, being poor, was obliged to give a bill of sale of his two cows and at length the men came after and drove them away. My mother and the children cried, but to no purpose. For the next six weeks we had nothing to eat but potatoes and salt.' "

Memories of the cold summer are enshrined in scores of New England and Canadian county histories. School children are still taken to view a tombstone standing in an abandoned cemetery at Ashland, New Hampshire. There one can read: "1771 Reuben Whitten 1847, son of a revolutionary soldier. A pioneer of this town. Cold season of 1816 raised 40 bushils of wheat on this land whitch kept his family and neighbours from starveation."

The remarkable forecast of 1816 in the Old Farmer's Almanac

There has been a legend, going back at least to 1846, concerning the *Old Farmer's Almanac* for 1816, then in its 24th year of publication. The story is that the publisher, Robert Thomas, had left a blank in his prediction for the first week of July, and that a mischievous printer, ordered to fill it in with anything that would fit the type-space, inserted "snow and ice." The uncanny prescience of this forecast was said to have made the reputation of the *Old Farmer's Almanac*. But there is no trace of this lucky prank in surviving copies.

CHAPTER 10

THE CHOLERA CONNECTION

A distant link to 1816?

CHOLERA OFTEN KILLS within a few hours of the appearance of the first symptoms: diarrhea, vomiting and fever. Rapid desiccation of the whole body ensues so that it becomes quickly wizened. The bursting of tiny capillaries in the skin causes the victim to turn blue or black. This is often swiftly followed by death. The rapid process of decay and dissolution is terribly frightening.

Origin of the pandemic

It has been suggested that one of the delayed consequences of Mount Tambora's eruption in 1815, and the crop failures and famine that followed in 1816–1817, was the first worldwide outbreak of Asiatic cholera (1816 to 1833).

Medical historians claim that before this great outbreak, cholera had been confined to the region of the Hindu pilgrimage on the Ganges, with sporadic incursions into China. They attribute the original outbreak that slowly spread to Europe and America to the famine in Bengal that followed the unusually cold summer of 1816, which many meteorologists attribute to Mount Tambora's eruption the year before. Could such a train of events be linked by cause and effect over a period of 17 years? In the spirit of today's environmental impact statement can we relate these disasters to one another?

A British lithograph of 1831 by Robert Seymour. Cholera by then was just crossing the Atlantic and beginning to ravage American cities. ILLUSTRATION COURTESY THE NATIONAL LIBRARY OF MEDICINE, LONDON.

Professor William McNeill suggests that the reason the 1817 cholera epidemic spread from India—where it had hitherto been confined to the area of the Hindu pilgrimage—was initially due to British Indian military operations, which had carried the disease to the Afghans and Nepalese. Once the infection was introduced into Asian trade routes and to the routes of the Moslem pilgrimage to Mecca and Medina, the disease reached from Morocco to the Philippines. Professor McNeill also suggests that there were, in fact, two waves of the epidemic between 1817 and 1832, but in any case the rate of spread seems to have been extraordinarily slow. In a sea voyage most of the infected would perish shortly after sailing and never reach the destination to infect others. On land the disease would spread more slowly than the speed that a man would walk. In an age before air travel or railroads, apparently it *is* possible the epidemic could take more than a decade to invade Europe.

The spread to Europe

By 1823 the cholera had spread to Tiflis and Baku and the shores of the Caspian Sea. Tsar Alexander I appointed a special medical board to combat it. When a renewed wave struck

A map showing the spread of the first great cholera pandemic.

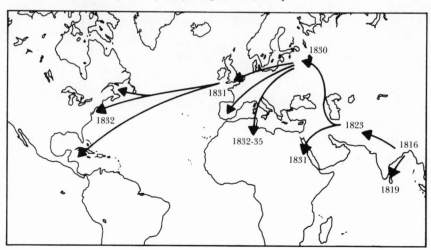

From early on in the 1830s cholera became a deadly fact of life in Europe. In this 1841 illustration by Honore Daumier in Fabre's Nemisis Medicale, *a cholera victim is ignored while casket bearers and carts carry off less recent corpses.*

Sevastapol in May–June, 1830, the brutal measures enforcing the quarantine led to bloody riots and the assassination of General Stolypin. After 50,000 people fled Moscow, Alexander Herzen compared the heroism of those who stayed to the spirit of 1812. Russian military campaigns then spread the disease to Poland and the Poles asserted that deliberate biological warfare was involved. In Pest more than 200,000 fatalities occurred.

Paris could not bury its dead, who were sewed up in sacks and stored in public buildings. More than 120,000 people fled the city. On March 29, 1932, there was a Mi-Caréme Ball—at the height of which people began to collapse—and within a few hours carriages were carrying away the dying. In May, Prime Minister Jean Paul Pierre Casimir-Périer himself succumbed, and the Soult Ministry was left to combat socialist uprisings in Paris and Lyons, sparked off by the misery and the social dislocations accompanying the plague.

The loss of 12 percent of the population of Cairo in 1831 led

In July of 1832, many in England thought the cholera epidemic was a fake. Six weeks later, the poster below was displayed in Dudley. REPRINTED COURTESY THE DUDLEY LIBRARY.

CHOLERA.

THE
DUDLEY BOARD OF HEALTH,
HEREBY GIVE NOTICE, THAT IN CONSEQUENCE OF THE
Church-yards at Dudley
Being so full, no one who has died of the CHOLERA will be permitted to be buried after *SUNDAY* next, (To-morrow) in either of the Burial Grounds of *St. Thomas's*, or *St. Edmund's*, in this Town.

All Persons who die from CHOLERA, must for the future be buried in the Church-yard at Netherton.

BOARD of HEALTH, DUDLEY.
September 1st, 1832.

q

W. MAURICE, PRINTER, HIGH STREET, DUDL.

Mehemet Ali to institute a board of health amongst the European consuls of the city. This served as a "cholera monitoring station" for years that could usefully alert the European powers of particularly virulent outbreaks during the Hadj. In Mecca, according to the traveller Dr. Edward Rüppell, in 1831, 50,000 people died in one fortnight.

Quarantine officials in the Canary Islands denied port entry to the HMS *Beagle* with the young Charles Darwin aboard on its outward-bound journey in January, 1832.

Arrival of cholera in New York, 1832

Alarmed by news of the events in Europe, New York City authorities instituted a wintertime quarantine in 1831–32. The populace eagerly read articles in the papers and magazines about the unfamiliar disease. There was a public lecture on cholera by medical quack Sylvester Graham, father of the "graham cracker," at the Baptist Meeting House in March, 1832.

Americans' fascination turned to dismay when news from Montreal confirmed the alarming fact that the scourge had crossed the Atlantic. On July 2, 1832, New York authorities were unable to conceal the appearance of cases of the disease within the city. Efforts at secrecy in order to prevent panic were of little avail.

The news leaked out.

By July 20 there were up to 100 deaths a day. Unburied corpses began to accumulate, the air was pungent with smoke from burning clothes and bedding. Empty houses tempted burglars and vandals. In August, churches were closed but there was no general breakdown in authority. Those in command believed that cholera's intended victims were among the lower orders of society. The *N.Y. Advertiser* spoke of a Mott Street prostitute who, "two and a half hours after decking herself at her glass, was carted away in a hearse." The broad social gap must have been almost as comforting an assurance of immunity as the expanse of the North Atlantic had been the month before.

The disproportionate mortality in the social stratum that "crowded the rats out of the vilest cellars" was so clearly a matter of sanitation and water supply that DeWitt Clinton proposed the Croton Aqueduct, the city's public pride for the next half century. Once again, a disaster was turned to ultimate good.

CHAPTER 11

OF LIGHTNING RODS AND SUNSPOTS

Does anybody really know what causes climate change?

I N 1816, MANY people associated the unusual summer cold with the occurrence of sunspots, which were clearly visible to the naked eye on the rising or setting sun that year, giving an unsettling feeling that something was amiss with the universe.

In earlier times, when Galileo first turned his telescope upon the sun and discovered these pox marks on its face, the fancied perfection of the celestial sphere was shattered for once and for all. But in the summer of 1816, the spots were considered to have some connection with the coldness. In a few extreme cases, the sunspots led to wild speculations about the end of the world, and to a French woman's suicide.

In Paris, an eminent philosopher, M. Ruoy, advertised that on July 12, 1816, he would give lectures on the spots on the sun, "in order to convince the credulous that there need be no fear of the extinction of that luminary, and consequently that the world is not speedily coming to an end; as reported by many malevolent and superstitious persons."

In America, the astronomical observatory at Williams College reported that nine groups of spots besides several single ones were scattered from the eastern to western side of the sun. Many amateur observers noted sunspots in their private dia-

Some in 1816 blamed Ben Franklin's lightning rods for the abnormally cold summer, saying these popular devices were interfering with electrical fields beneath the Earth's surface and thus altering the climate. Actually, Franklin (above) had shrewdly hit upon a far more reasonable explanation back in 1784, when he made the connection between climatic change and a dust veil resulting from the huge eruption of Mount Asama, Japan, the previous year.

ries. It is possible they were visible more easily than usual because of the slight darkening of the sun due to Mount Tambora's dust in the upper air.

Extravagant ideas

Another cause suggested at the time was related to the idea that the heat of the earth comes from heating due to electrical fluids circulating about beneath the surface. This was a period of intense interest in electrical phenomena. Although this effect today is estimated to be of negligible terrestrial importance, it has recently been advanced as the cause of the intense volcanic activity on Io, Jupiter's inner satellite. According to an article in the New York *Weekly Museum,* earthquakes that occurred in 1815 had interrupted these currents and the interior resistive heating reduced.

With all this electrical fluid flowing about, it isn't surprising that someone would strike upon the idea that the introduction of lightning rods was the culprit. And indeed this unlikely explanation was advanced at the time by a learned scholar at the Milan Observatory in Italy, who attributed the coldness of the season to the introduction of Dr. Ben Franklin's lightning rods. One wonders how many barns would have burned if an environmental impact statement had been required before they could be installed. But these ideas did contribute to the general merriment of the day, fanned by newpaper editors who had a great deal of fun making sport of them.

The noted acoustician E. F. F. Chladni in a German physical journal suggested that an outbreak of arctic icebergs had occurred—and cited some ship reports to support his idea. But this was a time when mariners were actually making observations of sea-surface temperature. There is nothing to support the idea that the North Atlantic was more than 1°F colder than usual during 1816. Evidence does not support a major aberration of the Gulf Stream during 1816, or even that the Gulf Stream had much to do with the period of generally lower temperatures that prevailed between 1780 and 1820, as has been proposed by one modern student of climate.

Franklin's shrewd guess

Unusually cold or warm seasons were noted by those with scientific curiosity even before 1816. At the end of the American Revolution the record shows an even lower world average yearly temperature between 1784 and 1786, apparently due to the great eruption of Mount Asama, Japan, in 1783—one of a series of natural disasters during the terrible famine of Temmei that darkened the last years of the Shogun Iyeharu, and in which a million people perished.

The extremely cold winter of 1783–1784 in the eastern United States is documented in David Ludlum's book *Early American Winters*. Among other phenomena, New York Harbor was blocked by ice for 10 days. Sleds could cross western Long Island Sound. The connection between the cold winter of 1783–1784 and high dust cover observed over Europe and the United States during the summer months of 1783 did not escape Franklin who, as early as May, 1784, wrote:

> "During several of the summer months of the year 1783, when the effects of the sun's rays to heat the earth in these northern regions should have been the greatest, there existed a constant fog over all Europe, and great part of North America. This fog was of a permanent nature: it was dry, and the rays of the sun seemed to have little effect toward dissipating it, as they easily do a moist fog arising from the water. They were indeed rendered so faint they would scarce kindle brown paper. Of course, their summer effect in heating the earth was exceedingly diminished.
>
> "Hence the surface was early frozen.
>
> "Hence the first snows remained on it unmelted, and received continual additions.
>
> "Hence perhaps the winter of 1783–1784 was more severe than any that happened for many years.
>
> "The cause of this universal fog is not yet ascertained. Whether it was adventitious to this earth, and merely a smoke proceeding from the consumption of fire of some of those great burning balls or globes which we happen to meet with in our course round the sun, and which are sometimes seen to kindle and be destroyed in passing our atmosphere,

a Table of thermometrical observations made at Monticello from Jan. 1. 1810. to Dec. 31. 1816.

	1810.		1811.		1812.		1813.		1814.		1815.		1816.		Mean of each month
	max.	min.	max.	min.	max.	min.	max.	min.	max.	min.	max.	min.	max.	min.	
Jan.	38	26	20	39	54½	34	13	35	16½	35	8½	35	16	34	36
Feb.	43	73		*	21	40	19	38	14	42	36	57	41	62	40
Mar.	20	61	28	44	06	78	31½	46	70	28	40	71	13½	43	73
Apr.	42	55	81	36	59	06	56	86	40	59	59	80	25	48	73
May.	43	64	88	46	62	79	39	60	06	46	62	81	35	59	82
June	53	70	87	56	73	99	58	74	92¼	54	75	93	57	69	87
July	60	75	89	60	76	89½	99	57	75	91	61	75	94½	60	89
Aug.	71	90	59	75	67	81	47	68	75	54	69	75	88	58	77
Sep.	55	70	81	50	67	81	39	55	80	32	53	70	52	70	61
Oct.	32	50	81	50	85	47	68	75	54	69	83	37	50	83	67
Nov.	27	44	69	32	35	62	18	43	76	20	40	71	23	47	71
Dec.	1A	32	62	20	30	49	13	35	63	18	37	53	19	38	59
Mean of each year	.55		.58		.55		.58		.58½		.58½		.54¾		

It is a common opinion that the climates of the several states of our union have undergone a sensible change since the dates of their first settlements: that the degrees both of cold & heat are moderated. the same opinion prevails as to Europe: & facts gleaned from history give reason to believe that, since the times of Augustus Caesar, the climate of Italy, for example, has changed regularly at the rate of $1°$ of Fahrenheit's thermometer for every century. may we not hope that the methods invented in latter times for measuring with accuracy the degrees of heat and cold, and the observations which have been & will be made and preserved, will at length ascertain this curious fact in physical history?

Thomas Jefferson was a much better statesman than he was a farmer. But he kept wonderfully detailed weather observations, as did many of the gentleman farmers of the day. COURTESY NATIONAL CLIMATE CENTER.

During the same 7. years there fell 622. rains, which gives 89. rains for every year, or 1. for every 4. days; and the average of the water falling in the year being 47½ I. gives .53 cents of an inch for each rain, or .93 cents for a week, on an average, being nearly an inch a week. were this to fall regularly, or nearly so, thro' the summer season, it would render our agri- -culture most prosperous, as experience has sometimes proved.

Of the 3905 observations made in the course of the 7. years 2776. were fair; by which I mean that the greater part of the sky was unclouded. this shews our proportion of fair to cloudy wea- -ther to be as 2776 : 1129 :: or as 5. to 2. equivalent to 5. fair days to the week. of the other 2. one may be more than half clouded, the other wholly so. we have then 5. of what astro- -nomers call 'observing days' in the week; and of course a chance of 5. to 2. of observing any astronomical phaenomenon which is to happen at any fixed point of time.

The snows at Monticello amounted to the depth in 1809. 10. of 16¼ I. and covered the ground 19. days.

10. 11.	31¾		31.
11. 12.	11.		11.
12. 13.	35.		22.
13. 14.	13¼		16.
14. 15.	29¾		39.
15. 16.	23.		29.
16. 17.	19¼		10.
average	22½		22.

which gives an average of 22½ I. a year, covering the ground 22. days, and a minimum of 11. I. and 11. days, & maximum of 35. I. and 39. days. according to mr Madison's tables, the average of snow, at his seat, in the winters from 1792.4. to 1801. 2. inclusive, was 23½. the minimum 10½ & maximum 38½ I. but I once (in 1772.) saw a snow here 3. f. deep.

and whose smoke might be attracted and retained by our earth, or whether it was the vast quantity of smoke, long continuing to issue during the summer from Hecla, in Iceland, and that other volcano which arose out of the sea near that island, which smoke might be spread by various winds over the northern part of the world, is yet uncertain.

"It seems, however, worthy of inquiry, whether other hard winters recorded in history, were preceded by similar permanent and widely-extended summer fogs. Because, if found to be so, men might from such fogs conjecture the probability of a succeeding hard winter, and of the damage to be expected by the breaking up of frozen rivers in the spring; and take such measures as are possible and practicable to secure themselves and effects from the mischiefs that attend the last."

Franklin speculated whether the dust he saw in the sky was due to breakup of meteorites or volcanic eruptions. He did not know, it appears, of the disaster in the hermit kingdom of Japan. In his time the theory and evidence about the Great Ice Ages of prehistory were still unexplored. But he seems to have had a very shrewd and clear idea of how year to year changes of world temperature might be caused.

If anything characterizes the early explanations of the cold summer of 1816 it is the paucity of data rather than the lack of ideas. Theories were a dime a dozen in climatology, and they often are not worth much more today.

The main task of climatology has always been documentation rather than explanation. Our American forebears knew this, and that is why they busied themselves with maintaining the college meteorological network. Between 1790 and 1810 the scientific literature was full of articles concerned with climate change. There were studies of possible change since the time of first settlement. Explanations, such as deforestation, were offered.

Noah Webster published a long paper in one of the first issues of the Transactions of the Connecticut Academy of Sciences in which he compared contemporary European weather with that of Roman times.

Many others with a classical education wrote on this theme. Their dependence upon the casual meteorological remarks of the classical authors convinced them of how desperately quantitative data was needed. It persuaded them to collect data systematically of their own, and thus provide the documentary evidence—as pathetically thin and incomplete as it is—of the climate of their time. It was a wonderful statement of their faith in the future to deliberately bequeath to us this information.

The beginnings of the temperature record

The 1816 data is limited to measurements of temperature, barometric pressure and estimates of cloudiness and winds at points in northeastern America and western Europe. The rest of the world was practically unmonitored. Even today vast oceanic areas are poorly covered with surface measurements, and satellite-determined surface temperatures are not quite good enough. The 1816 data is only at the surface. We do not know what transpired in the upper atmosphere. A very few sea captains—such as John Hamilton on the Liverpool-to-London packet—bothered to observe air and sea temperature on a regular basis. Observations of sea temperature beneath the ocean surface were almost nonexistent: The only ones we know of in 1816 were a few obtained by the Russian explorer Otto Kotzebue off Hawaii.

From these sparse beginnings the geographical coverage of surface measurements gradually increased during the 19th century. Upper air measurements were not gathered systematically and globally until after World War II. It is therefore difficult to judge how meaningful estimates of "average world temperature" before the 1880s really are. Monitoring of changing deep-ocean temperatures is very sparse even today. It is a rash oceanographer who will announce any statement as to whether the ocean is warming or cooling as a whole. It would be nice to know, because the ocean stores a great reservoir of heat that could affect climate.

We can scarcely dispute the notion that the primary task of

Beginning in 1812, Samuel Rodman Jr. started keeping meteorological records at his home in New Bedford, Massachusetts, that would be kept up by his descendants until 1905. His records, sparse as they are, were among the most detailed of any kept during the cold season of 1816. COURTESY NATIONAL CLIMATE CENTER.

those interested in climate is to discover new kinds of data, new techniques of analysis and to assemble measurements with the greatest geographical coverage, at all altitudes in the air and depths in the ocean, as frequently as possible. Every hopeful lead, be it in tree rings, glaciers, ice cores or the ocean bottom, needs exploration.

In contrast to the unifying role that theory plays in the exact sciences, theory in climatology is a *gloss* that keeps the task interesting.

To a large extent, science rests upon numbers. Temperature was one of the first meteorological variables that was recorded routinely. By 1816 the mercurial and spirit thermometer (sealed in a glass tube) was a reliable instrument. It had a history of 150 years. Prototype sealed thermometers were constructed in Florence in the mid-1600s by a master-glassblower at the request of Ferdinand de Medici, the Grand Duke of Tuscany. Miraculously some still survive in Florence's Museum of the History of Science. The Duke wanted a reliable ther-

June 1816.

Saturday 1. S.W.   57, 82, 71 Fair.

Sunday 2.   S.E.   66, 77, 56 Fair.   a.m.

Monday 3.   N.W.   56, 66, 56 Fair. Moon quartered 3h. 46 m

Tuesday 4   S.E.   46, 64, 51 Cloudy. Uffing to Concord. Little rain.

Wednesday 5. W.   60, 83, 71 Fair. thundershower rain 2 10ths - some thunder & lightning at Uffing. At Concord.

Thursday 6. N.W. & high 61, 50, 39 Cloudy. A little snow fell at Concord.

Friday 7. N.W. & stong 38, 49, 43 Fair. At Concord.

Saturday 8. N.W. & stong 39, 45, 40 Fair. little snow at Concord.

Sunday 9. N.W. 42, 60, 41. Fair. Moon totally eclipsed in the evening - At Concord water froze in the night.

Monday 10. N.W. 41. 53. 40. Fair. At Concord water froze in the night 1/4 of an inch thick. For the last 5 days fire has been necessary in the parlor & in the council chamber.

Tuesday 11. W. 43, 54, 46 Fair - a small shower; at Concord.

Wednesday 12. N.W. 48; 70. 64. Fair. At Concord in the morning thunder, lightning & rain.

Thursday 13. N.W. 57, 75, 61. Cloudy, foggy at Concord. On the 6th instant there fell snow all day - much of it dissolved as it fell but next morning it lay 2 inches thick on the ground on a level - this was at William=stown in Vermont.

Friday 14. S.W. 52, 61, 53. Cloudy, thunder lightning rain 4. 10th At Concord.

Saturday 15. N.E. 51, 61, 55 Cloudy. rain 4 10th At Concord.

Sunday 16. N.E. 54, 64, 49 Cloudy. At Concord.

NE. 49, 61, 55. Cloudy. rain 1. 10th At Concord.

51, 58, 58. Cloudy. rain 1, inch 10th. At Concord.

Governor William Plumer's New Hampshire weather journal. **COURTESY NATIONAL CLIMATE CENTER.**

Wednesday,   June 1816.

N.W. 62, 79, 71 Fair. At Concord.

N.W. 56, 65, 58. Fair. At Concord.

Friday 21. W. 53. 79. 68. Fair. At Concord.

Saturday 22. S.W. 90, Fair. At Concord.

Sunday 23. S. 72, 93. 84 Fair - At 4 O'Clock P.M. thermometer at 100. - At Concord.

Monday 24. S.W. 78, 94. 83. Fair. At Concord, thunder lightning & little rain.

Tuesday 25. N.E. 69, 58, 54 Cloudy, rain - At Concord. Moon changed 9 h. 21 m. A.M.

Wednesday 26. S.W. 50, 68, 57. Fair. At Concord.

Thursday 27. N.E. 50, 56, 58. Cloudy - Rain, At Concord.

Friday 28. N.W. 62, 76, 62 Fair. At Concord.

Saturday 29. N.W. 52. 66, 53. Fair. At Concord.

Sunday 30 W. 50, 68, 54 Fair. At Concord frost in night.

Summary for June

The thermometer at the highest was 94 at the lowest 38, it averaged 59 2/3 & little more

The wind was west 4 days, south west 6, south east 2, north east 6, North west 12 days.

It was fair 20 days, cloudy 10; rained in 11 days, thunder 4, lightning 4, Snow 2 & fogg in one day.

Of rain there fell 3 inches 1. 10th

Note. On the 22d thermometer at Salem M 93; on the 23d 101; 24th 100.

On the 22d from Boston 7 spots were seen in the Sun.

At Waltham M on the 10th at sunrise thermometer at 33 water froze; on the 23d at 3 o'clock P.M. at 99.

This month was cool in Europe - In France, Italy &c fire was comfortable - on the 6th & 9th falls of snow in England.

mometer for determining whether the water in deep wells, which felt warmer in winter, changed temperature during the year.

The first attempt to systematize the collection of meteorological temperatures was made in 1723 when James Jarin, the Royal Society's secretary, distributed standard thermometers to the learned of the civilized world with the request for regular

An example of an early thermometer, this one a spiral instrument on display at the Museum of the History of Science in Florence. It is divided into 420 degrees. While, in theory, it was possible to make highly accurate thermometers, in practice their accuracy depended on the talents of the glass blowers who actually made the instruments.

temperature data. It was one of these that hung in the south-west room of the Harvard president's house and registered the heat wave on June 18, 1749. These thermometers were of the upside-down James Hauksbee scale, and unfortunately tended to differ from one another.

A great variety of scales existed. One surviving thermometer in Holland has 18 scales engraved upon it. For example, there was a Fowler scale for use by horticulturists in monitoring their hothouses. The famous botanist Carl (von Linne) Linnaeus is said to have been the founder of the Celsius scale, having turned upright an upside-down scale advocated by Anders Celsius.

It was a Hauksbee thermometer that was being used for regular temperature records at the astronomical observatory in Uppsala, Sweden, in 1731. Celsius himself was doing the observing. At the same time René-Antoine Ferchault de Réaumur was introducing his own thermometer—and his scale was to survive as the most popular in eastern Europe for many years. The centigrade scale (Celsius) became official in France on April Fool's Day, 1794, with the revolution.

The young Danziger Daniel Gabriel Fahrenheit may have been influenced by Galen's ancient "eight degrees of heat," which was in use by the medical community of his time. His choice of a binary scale of temperature was published in the Philosophical Transactions of the Royal Society in 1724. Contrary to general belief, it was calibrated at three temperatures: 0°F, the temperature of a snow and sal ammoniac mixture; 32°F, the temperature of melting ice; and 96°F, "body" temperature. Evidently the use of boiling water for the 212°F point was not employed.

By the 1780s, methods of accurate calibration and standardization had been developed by Henry Cavendish. The long-term stability of large numbers of mercury thermometers was studied by the famed Antoine Lavoisier himself in the constant temperature cellars 210 weary steps down beneath the Paris Observatory. By 1816, there existed a network of routine meteorological observing stations with suitable meteorological instruments capable of documenting the anomalously cold summer to come.

Indirect indicators of temperature

To extend the data base to the years before direct thermo-metric temperature measurement, climatologists use many other variables: width of tree rings, dates of harvest, bands in coral, foraminifera in deep-sea sediments and length of gla-ciers. It is not a trivial matter to extract unambiguous estimates of temperature from such variables. Fluctuations in the width of tree rings can be caused by time of year and by rainfall as well as by temperature. The numerical formula that relates these variables to temperature is imperfectly known. Each of these variables contains some information about past tempera-tures in complex ways. Disentangling these relationships is a serious task. There is no assurance of complete success. It is like an immense, multidimensional, many-colored Rubik's hy-percube that must be manipulated until meaningful patterns emerge.

Reconstructions of global temperature extending over many centuries are based on indirectly determined temperature esti-mates. Only in the last 100 years can good global averages be assembled from world-wide surface data. The two sets of data are not strictly comparable; splicing them is risky.

A multitude of periods

Long-term records show variations with many apparent pe-riods greater than the expected annual variation. It is plausible to assume that what causes short-period variations may differ from what causes long-period variations. The sun's radiation is the main source of our earth's heat, so it is also plausible to first seek a cause in variation in the sun's heat that reaches the earth's surface. The variation in sunlight could occur at the sun itself due to incompletely understood processes at work within it. Or, it could come about because varying amounts of dust in the upper atmosphere of earth reflect sunlight before it pene-trates the lower atmosphere. Many of today's climatologists tend to prefer the former cause as an explanation for long-period climate changes, and the latter cause to explain fluctuations of only a few years.

W. J. Humphreys, volcano theory advocate

One of the foremost advocates of the theory that cold years are caused by less radiation from the sun reaching the earth—both by sunspots and volcanic dust suspended high in the atmosphere—was W. J. Humphreys of the U.S. Weather Bureau. His wartime lectures on meteorology to aviation cadets at San Diego beginning in 1914 were published in a famous textbook, *Physics of the Air* in 1920.

In it, Humphreys assembled the historical record of sunspots observed between 1750 and 1913. Also, a synthesis of temperatures from many stations that could be regarded as departures of world temperature from the normal. It also tabulates the time and size of volcanic eruptions. Humphreys concluded: "It appears that the dust in our own atmosphere, and not the condition of the sun, is an important, if not the controlling, factor in determining the magnitudes and times of occurrence of great and abrupt change of insolation intensity at the surface of the earth."

In Humphrey's analysis the cold year of 1816 is clearly depicted as a result of the eruption of Mount Tambora. As we know, the low temperatures occurred mostly during the extraordinary summer itself.

Much more research on the causes of climate change has been done since Humphrey's lectures of half a century ago. In recent years increased use of computers, a more sophisticated understanding of statistics, the growing number of scientists, longer instrumental records and careful reconstructions of the past have all contributed to clarifying possibilities, if not to defining answers.

The main feature of the global temperatures of the 19th century is the slow rise of temperature from the cold beginning to a warm end. No one supposes that this trend is the result of a particular volcano. In fact, many present-day climatologists are inclined to believe this reflects a general rise of radiation from the sun itself. Superimposed upon this trend are fluctuations of only a few years' duration, many of which have been attributed to the dust thrown up by individual well-known volcanic eruptions. The cold summer of 1816 shows up on these curves of estimated world temperature as a dip in temperature on a curve that is already quite low for the period 1780 to 1820.

Complexity of the temperature record

To see how this statement is supported by data we can inspect a graph made up by the climatologist J. Murray Mitchell. It displays yearly temperature averages of many surface stations distributed widely over the northern hemisphere. The geographical coverage of these stations is not perfectly uniform, but the results are supposed to be representative of what one would find if one made a truly uniform average over the northern half of the world. The series does not begin before 1880 because there simply were not enough stations distributed over the hemisphere to ensure that averages are not too heavily weighted toward North America and Europe. This graph shows that the warming trend in the 19th century continues well into the 20th century, until about 1938. Some of the dips in temperature with duration of several years can be associated with individual volcanic eruptions, as we will shortly show.

When we try to extend estimates of world temperature to years before 1880 we rely more on arguments from individual stations. Here we encounter the fact that year-to-year variability of individual stations is largely due to many local random events. It is accordingly difficult to discern smaller long-period fluctuations. Inspection of the mean June temperatures at New Haven (p. 38) will illustrate this problem. Long-period changes are obscured by a welter of yearly ups and downs. To filter out these rapid changes, climatologists who study ancient records smooth them by plotting running averages rather than values of individual years. This process enhances the relative amplitude of the long periods. But it also tends to wipe out periods associated with cooling by volcanic dust.

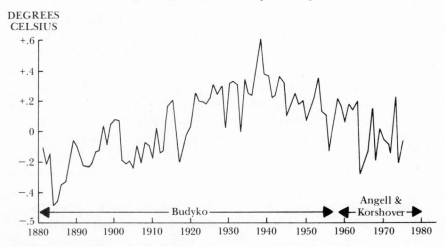

Recorded annual mean temperature of the northern hemisphere (Murray), assembled during a period in which substantial geographical reporting coverage existed.

Graph showing 30-year running averages of temperature in Holland from 1735 to 1944 (from Labrijin).

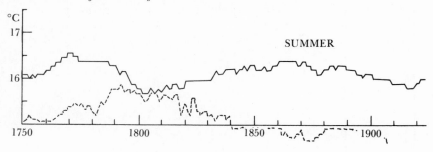

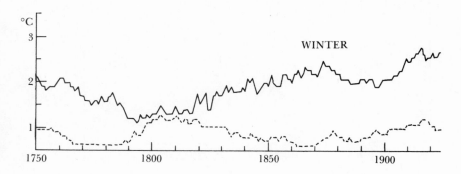

There are a few records of thermometer readings in Europe that extend back into the early 18th century. Those from Holland and England are particularly useful. The graphs present overlapping 30-year averages. That is pretty heavy smoothing. Both the records indicate a warming trend beginning about 1790 in winter. The summer record does not show as strong a trend. The cold period from 1780 to 1820 is clearly visible.

It would take us too far afield to pursue obvious questions about how much of this warming may be due to local urban heating at the points of observation. There are many possible pitfalls and climatologists try to detect them by comparative studies of different individual station data. Sometimes stations are moved within a city, sometimes from downtown to the airport, often the exposure of a thermometer is changed, or the hours of observation. Constructing a reasonably reliable long thermometric record is a troublesome and difficult task.

A little ice age

The 40-year-long cold spell that embraced the cold summer of 1816 is one of a series of similar events that occurred in historical but prethermometric times. It was during this 40-

Running mean temperature in Lancashire, England, from 1740 to 1940.

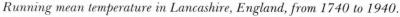

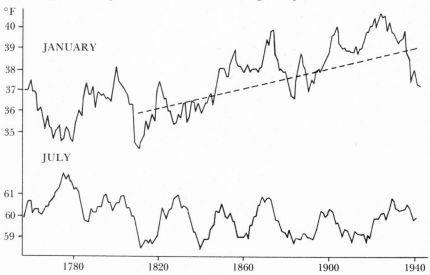

year period that the glaciers of northeast Greenland reached their maximum extension, a fact supposed to be associated with the disappearance of the North Greenland Eskimo. The earlier temperature minimums beginning in the 14th century have been called "little ice ages". Their most dramatic effects were upon Iceland, where farms were covered with permanent ice fields, the remains of which are only now being uncovered as the ice retreats. In West Greenland the glaciers advanced over a region where there was once a settlement of 300 Norse farm-steads, 3,000 people and a large number of sheep and cattle.

Effects of sunspots

Dr. J. A. Eddy of the Smithsonian Astrophysical Observatory has advanced evidence to show that these major long-period events are related to changes in solar radiation. His demonstration hangs upon a novel interpretation of the meaning of sunspot numbers. He has compiled the most reliable telescopic sunspot numbers going back to 1610 A.D. Because they extend back so far and because a single number holds for the whole earth, extensive geographical coverage is not required. Thus sunspot number statistics have been very popular. The most striking feature of sunspot numbers is a strong 11-year period. Some meteorologists have exhausted their ingenuity and energy trying to squeeze out of meteorological measurements an 11-year pattern to compare with the wonderfully strong sunspot cycle. Their lack of success, their persistence in seeing relations that less committed observers could not verify, and their obvious emotional involvement all contributed to a general slackening in the belief that sunspots have anything much to do with weather.

It was Eddy's inspiration to assert that the number of sunspots has little, if any, effect upon solar radiation, but that long-period fluctuations in solar activity *are* associated with the amplitude of the 11-year cycle. He drew a curve that touched the tops of the successive sunspot maxima as a measure of the strength of solar radiation leaving the sun. The greatest dip in this curve occurs between 1640 and 1715 and is called the

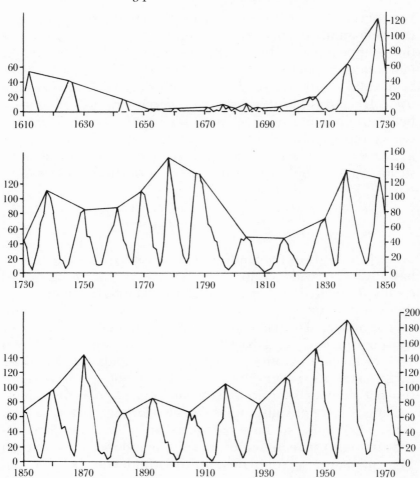

The solid line forms an envelope over peak sunspot periods between 1610 and 1975 (Eddy), showing long-range change in sunspot activity.

Maunder Minimum, a period known to be excessively cold, but occurring before the time of thermometric records. The second cold period, 1795 to 1825, includes the cold season following the Tambora eruption.

Eddy has gone further by comparing measurements of atmospheric radiocarbon in tree rings with temperature during the historical period and extending his solar activity curve back —by this means—to before 5,000 B.C. Using this extended curve he is able to resolve the time of the Spörer Minimum

(1400–1510 A.D.)—which was so hard on Iceland—and the warm time of the Medieval Maximum (1120–1280 A.D.), during which the kindly climate enticed the Norsemen to settle there.

Now, of course, an inspired guess like this does not constitute a formal solution of the inverse problem. But it points the way. Let us accept it—until someone tells us better—as the explanation for the several cold decades in which the cold summer of 1816 is embedded.

Spontaneous random fluctuations of climate

One reason for being cautious is that numerical models run on computers have shown us that even in the absence of any external fluctuation in solar radiation received by the earth the model itself fluctuates spontaneously enough to produce noticeable long-term changes in globally averaged temperature.

A recent example of such a simulation run on a computer has been made by Professor Alan Robock of the University of Maryland. The numerical model of the atmosphere that he uses is one that budgets at each latitude the heat flux across latitude circles by atmospheric eddies (or weather systems) as well as the vertical flux of heat through the top of the atmosphere.

The north-south exchange of heat by the atmospheric eddies in the model is a random one, whose magnitude and period is adjusted to be representative of that actually occurring in the atmosphere. It has the same statistics as that of the real atmosphere, but there is not a one-to-one correspondence of the model's eddies with any real ones. Robock's first simulations of climate change were made in the absence of any external forces —either dust or sunspots. Even under these conditions there were long-period changes in the world temperature of the model, which seem qualitatively very much like those observed —statistically speaking, because there is no way to attach specific dates to any of the model's random predictions of climate change.

But the model tells us to expect that the temperature of the

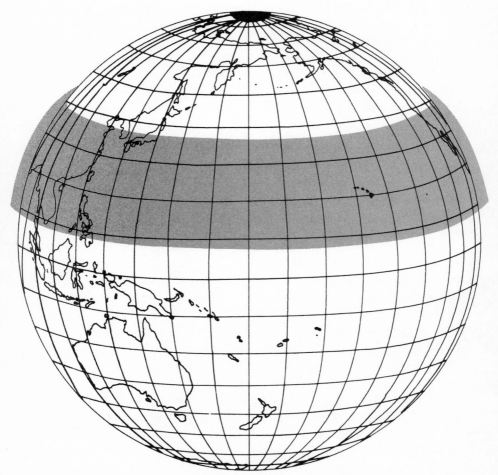

In April of 1982, aerosols and dust from the eruption of El Chichon, Mexico, circled the northern hemisphere. The cloud is believed to be 140 times denser than that thrown up by Mount St. Helens in 1980.

real atmosphere, in the absence of any exterior forces, could spontaneously wander up and down over several hundred years within a range of about 1.5°C, purely by spontaneous and unpredictable internal redistributions of heat by atmospheric eddies. Each time the model was run a curve was produced, different from the last in every particular detail, but with the same overall characteristics. To simulate the actually observed temperature over past centuries by this mechanism would require detailed information about the heat transport by the eddies, and this we do not have. Although this random model does not enable us to compute specific, dated past world temperatures, it does warn us that much of the past observed variability may be thought of as due to random processes within the atmospheric circulation, and that we should not expect to obtain perfect correlation of past temperatures with any model that includes external forces—such as sunspots and volcanos.

Having examined the magnitude of climatic change due to internal random processes, Robock then turns to testing the effects of sunspots and dust. Despite Humphrey's gallant attempt to make such a table some 60 years ago, it is clear that the sparcity of early data makes it difficult to compute world average temperatures much more than 100 years ago. The uneven distribution of weather observing stations in the southern hemisphere is a serious limiting factor even today. Robock uses a new compilation of past world temperatures in the northern hemisphere. It does not go back beyond 1880, and different authors offer different estimates.

Factoring in volcanic dust

As an input to his model, Robock needs data on the number of sunspots each year—one of the better observed sets of data —and also a measure of the amount of dust in the atmosphere. This last is not, unfortunately, based on direct measurements of dust, but estimated from the size of each volcanic eruption. The British climatologist H. H. Lamb has assembled a historical listing of these rather arbitrary estimates of dust injected into the upper air. Robock used Lamb's "dust veil index."

In order to use these data in a numerical model it is necessary to know how to convert the information on the number of sunspots and from the dust veil index into an estimate of the actual changes of incoming solar radiation. Unfortunately, there are some difficulties with this. Some solar physicists now claim that the relation between the number of sunspots and the incoming radiation is not a simple proportion, and there is enough uncertainty about the exact form of the relationship for Robock to try several different laws. But he could find no significant correlation between the number of sunspots as the external force affecting climate and the observed world temperature. As a matter of fact, the most casual inspection of the record of past world temperature reveals no evidence of an 11-year cycle—the most striking feature of the sunspot numbers.

The same inspection, however, does show what appear to be sudden drops in temperature following some major volcanic eruptions.

Making this quantitative model is a little difficult because methods of determining the dust veil index are crude and full information about volcanos is not always available.

Nevertheless, if the dust veil index is used to compute a decrease in incoming solar radiation in direct proportion to the amount of material suspended in the atmosphere, a surprisingly good correlation of the computed model temperature with the observed world temperature results. Robock's comparison does not extend back to 1816 when the world temperature was so sparsely measured, but it does suggest strongly that subsequent eruptions such as that of Krakatoa, Katami and Agung actually depressed world temperatures.

The volcanic "signal"

At the National Center for Atmospheric Research in the foothills of the Rockies overlooking Boulder, Colorado, in 1975, using a simple mathematical model of overall global radiation balance, Stephen Schneider and Clifford Mass computed the historical global surface temperature. They used as input data new tabulations of sunspot numbers and new esti-mates of density of volcanic dust. The computing machine pro-duced a table of global temperatures. It shows most strikingly the wonderful low-temperature decades of the early 1800s and even our special year 1816.

They found, in general, an abrupt fall in global temperature approximately one year following isolated strong eruptions—amounting to about 0.3°C—and a slow recovery of the temper-ature over several years. They conclude—with the cautious re-straint of scientists—that there is "a definite, albeit weak, volcanic signal identifiable" in the temperature records, and that dust is most probably an important contributing cause to climatic variability, but certainly not the only cause.

Because the events surrounding the cold summer of 1816 are not adequately documented, it is important to test, by studying more recent eruptions, the idea that temperature drops after large volcanic eruptions. In modern times we can explore any effects on the globe as a whole in different latitude bands and at different altitudes.

A modern case history: Mount Agung's 1963 eruption

Two meteorologists, J. K. Angell and J. Korshover of the National Oceanic and Atmospheric Administration, published in 1977 a study of the global change in temperature in the years following the eruption in the spring of 1963 of Mount Agung in Bali—a geographical position very much equivalent to that of Mount Tambora. Although the global change was a drop of only 0.2°C, the effect was more intense in northern extratropi-cal latitudes, where it amounted to a decrease of 0.6°C.

Since the dust veil index of Mount Tambora has been esti-mated to have been perhaps four times larger than that of Mount Agung, it seems plausible that northern extratropical latitudes could have experienced temperature drops of as much as 2.4°C during the two years immediately following the eruption. This is an effect very much like the temperature anomalies actually observed at New Haven and Geneva during the summer of 1816. Robock's numerical model also shows a greater sensitivity in the northern hemisphere—and even over land as compared to over sea.

The changeling climate

It would simplify life considerably if anomalies in weather were uniform over large portions of the globe—and if geographically uniform causes always resulted in geographically uniform response. In past scientific literature, concern has been expressed that, even with the limited data available, it is clear that 1816 was not the absolutely coldest year everywhere.

For example, U.S. Weather Bureau scientists H. E. Landberg and J. A. Albert have searched temperature records from various European stations to see how 1816 ranks as a cold summer elsewhere than in New England. Defining a summer as June through August, 1816 is the coldest summer of the 187 years of observations at New Haven and Geneva, and the second coldest in the 234 years observed at Philadelphia. In Rome, Edinburgh and Budapest, 1816 ranks 6th through 8th, and there is nothing particularly exceptional at all about it at Copenhagen, Wilno and Vienna.

Undoubtedly there was local variation in the magnitude of the observed temperature anomaly, a fact that is in accord with the idea that in the absence of a completely uniform response everywhere, some local responses will be hotter, some colder than the average of the overall cold anomaly. The idea of ranking years according to "coldest ever recorded" can be quite misleading.

Trans-Atlantic links

It is interesting that the cold events in New England were linked with cold events in Western Europe. But this is a very normal thing to have happened, according to American meteorologist Jerome Namias. In his work correlating meteorological variables between different parts of the globe, he has found a very strong coupling of summertime pressure and circulation between New England and western Europe. Thus such a linking is to be expected.

Finally, Robock's model demonstrates how grateful we ought to be that the dust from Mount Tambora fell out from the atmosphere after a few years. If it had remained there indefinitely, continuing to depress the amount of incoming radiation by a steady 2 percent, the world would be a lot colder now, and by the year 2315 A.D. the average global temperatures might have dropped by 10°C. That is just about what it would take to produce another bona fide ice age. But we were saved by the fallout.

New kinds of data

Scientists have not exhausted all possible sources of data that can help them determine Mount Tambora's role in the climatic disturbances of 1816. The volcano itself has not been properly investigated by modern volcanologists. However, it already appears that the stimulus given to volcanology and climatology by the eruptions of Mount St. Helens and El Chichón and speculation about its climatic consequences, if any, has focussed attention again on Mount Tambora in particular, and the possible link between volcanos and climate in general. New ideas and facts will emerge. For example, in November, 1980, the U.S. National Aeronautics and Space Administration sponsored a "Symposium on Mt. St. Helens' Eruption." The conference report, published in *Science* magazine (23 January 1981), states that instead of dust in the upper air being the agent that lowers incoming sunlight, aerosols composed of minute droplets of sulphuric acid may be the real culprits.

If so, we are in an excellent position to read the record of the past. Recently, Dr. C. U. Hammer and colleagues of the University of Copenhagen have been studying very long cores obtained by drilling into the top of the glacier that covers Greenland. Some of these cores are as long as 1,390 meters and carry embedded in them undisturbed layers of volcanic debris and sulphuric acid dating back 10,000 years—a record that spans virtually all history, even back to the fabulous explosion in the Aegean Sea that gave rise to the myth of sunken Atlantis. The cores are extremely detailed and can be dated in the distant past to within a year or two. It appears that the record of acidity is clearly related to past volcanic activity.

On the basis of these cores, the Danish scientists estimate that in 1815 Tambora threw seven times more sulphuric acid aerosol into the upper air than its Indonesian neighbor Mount Agung threw up in 1963. Since most climatologists agree that Mount Agung depressed world-wide temperatures by 0.2°C for several years following its eruption, it is not unreasonable to argue that the acid thrown up by Mount Tambora depressed world temperature by 1.4°C during 1816 and local effects could have been even more severe.

Mount St. Helens and the weather

Finally, to bring the subject up to date, we now have available the first climatic tabulations of the northern hemisphere surface temperatures during 1980 and 1981. In a very careful study by three scientists of the University of East Anglia, England—Drs. P. D. Jones, T. Wigley and P. M. Kelly—reported in the February, 1982, issue of the *Monthly Weather Review* that there is no evidence of any cooling attributable to Mount St. Helens' 1980 eruption. Evidently it was too insignificant to have any noticeable effect. The tabulation shows that 1980 was only 0.08°C warmer than 1979, and that 1981 was a whopping 0.48°C warmer. In fact, 1981 emerges as the warmest northern hemisphere surface temperature for the historical record 1881–1981.

The promising 1982 eruption of El Chichón

The dust and sulphuric acid droplets thrown up by the eruption of the Mexican volcano El Chichón on April 4, 1982, may be quite another matter, however. The dust cloud was first observed at the National Oceanic and Atmospheric Agency's observatory on the top of Mauna Loa, Hawaii, on April 9. By April 23 it extended from an altitude of 18 to 27 kilometers, gradually spreading in latitude.

It is perhaps premature to record here the numerical comparisons of the intensity of this cloud compared to that of Mount St. Helens, but it is a great many times denser. So much denser, in fact, that meteorologists, such as Brian Toon of the National Aeronautics and Space Administration, are being quoted in the press as predicting measurable cooling for the next year or two.

Very reliable natural scientists, who have spent years roaming the peaks of the Hawaiian mountains, tell us that the effects of the dust cloud are clearly visible to the naked eye. Ordinarily, the air at these altitudes is so clear that one can exclude the direct rays of the sun from one's eyes by simply extending a hand upwards to cover it, and that previously the sky was always a dark blue. Now, they report, the light is so scattered by dust that one cannot block the sun's light so easily, and the sky is noticeably grey. Confirmation of any climatic effects which this impressive cloud of dust and sulphuric acid droplets may have upon northern hemisphere surface temperature must await tabulation of the observed temperatures all over the world.

CHAPTER 12

A PERSONAL LOOK
AT THE RECORD

You be the judge

THE QUESTION OF whether volcanos cause two or three-year periods of cooling over the world after their eruption is so complicated—and the early data is so sparse—that it is easy to understand why scientists cannot agree. The reader is invited to judge for himself.

For this purpose two graphs are presented. The first is for a 50-year period, 1780 to 1830, embracing Mount Tambora's eruption and 1816, "the year without a summer." The second graph is for the years 1881 to 1981, in which the historical meteorological data is sufficiently numerous and well distributed geographically to provide monthly and annual estimates of mean surface temperatures for the northern hemisphere.

In the first graph temperatures are shown for three selected stations: June temperatures for New Haven, Connecticut; summer temperatures for Geneva, Switzerland; and summer temperatures in central England. Some other records could be displayed, but there are not enough to make meaningful estimates of the mean temperature for the whole northern hemisphere. During 1780 to 1830 continuous weather records were gathered in eastern North America and western Europe, but nowhere else. Below the temperatures the time of various volcanic explosions is indicated. The height of the volcano is drawn to distinguish large volcanic explosions from smaller ones. The sizes given are what we call "volcanic rank" (the Smithsonian's volcanic explosivity index subtracted from 8). Eruptions smaller than rank 4 are not shown. This size is probably not proportional to the amount of dust and acid droplets thrown up into the stratosphere, but it has the advantage of being uninfluenced by previous knowledge of climatic data— as Lamb's dust veil index is said to be. It also does not distinguish the latitude of the eruption or the chemical composition of the material ejected, which may be important. There is some uncertainty of the date, around 1800, of Mount St. Helens' earlier eruption.

At the top of the figure we show the annual mean sunspot numbers scaled from Eddy's graphs as determined from his research at the Harvard College Observatory.

The main feature of all the three temperature curves is that they are wiggly, wandering irregularly up and down a few degrees Fahrenheit. This must be attributed mostly to random local variations in weather. The feature that they have in common is the dip of from 3°F to 6°F in 1816, the year following the singularly large eruption of Mount Tambora. The marks indicating the years are centered at mid-year in summer. That explains why the eruption of Mount Tambora is placed just earlier than the mark for 1815 (it was in the spring of 1815).

All the other eruptions were smaller than that of Mount Tambora (this is true for the past 10,000 years). Whatever effect these other volcanos may have had on summer temperatures of these individual weather stations escapes the eye. It is difficult to convince oneself that any summer cooling followed the eruption of Mount Asama in Japan and Mount Lakagigar in Iceland in 1783–1784 from these temperature records. There is too much wandering up and down of the curves to be sure that one is not fooling himself. You be the judge: do you see a temperature drop after Mount Tambora, or after any of the other volcanos?

The curves of sunspot number at the top look rather like ocean waves, with a period of 11 to 12 years. Is there any obvious similarity between them and the temperature curves? If you will grant that 1816 was noticeably cold, was that because there were more sunspots than usual, or less?

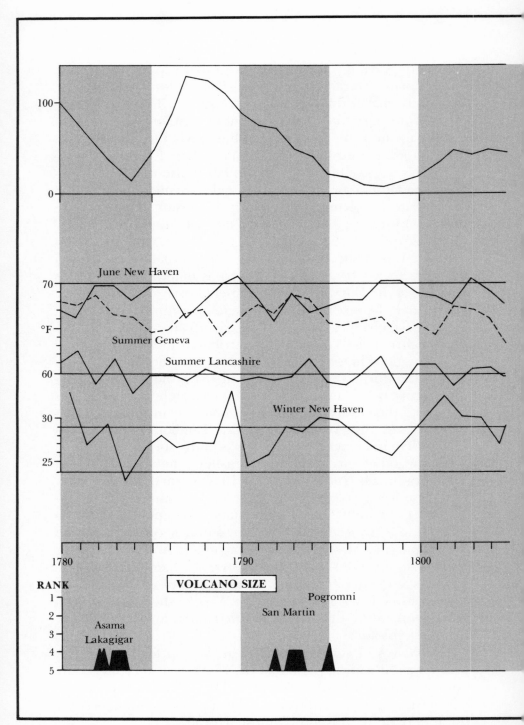

100

70
June New Haven

°F
Summer Geneva

60
Summer Lancashire

30
Winter New Haven

25

1780   1790   1800

RANK   VOLCANO SIZE
1
2   Pogromni
3   Asama   San Martin
4   Lakagigar
5

Judge for yourself whether there is a relationship between sunspot activity and climate change or, likewise, between volcanic activity and climatic variation. This chart records

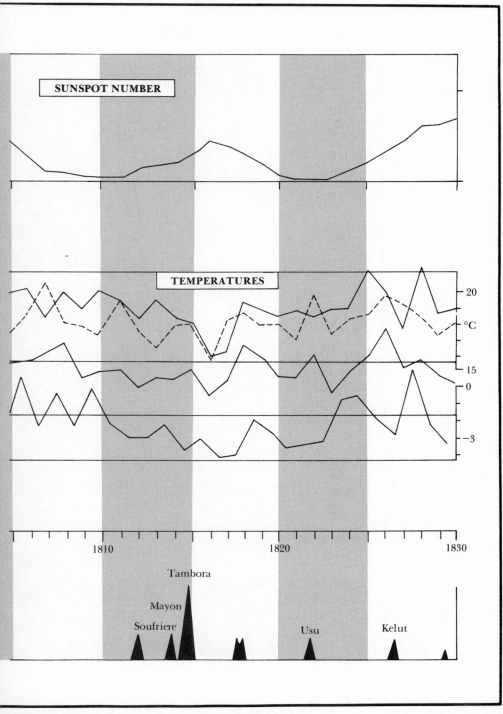

weather temperature changes between 1780 and 1830 in Geneva, Lancashire and New Haven—three places where accurate meteorological records were being kept.

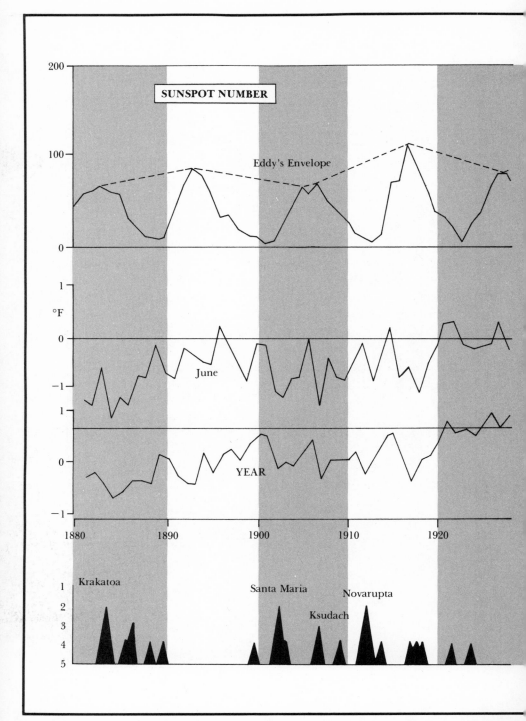

As in the previous graph, this illustration allows you to judge for yourself the relationship between sunspots, volcanos and weather. Note that in this graphic Eddy's envelope of peak sunspot activity is included. And temperature readings are averaged for the northern

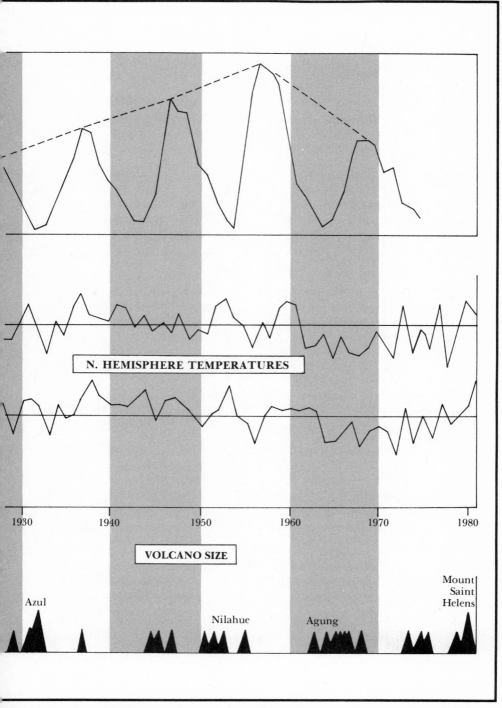

N. HEMISPHERE TEMPERATURES

VOLCANO SIZE

1930   1940   1950   1960   1970   1980

Mount
Saint
Helens

Azul

Nilahue   Agung

hemisphere, since during the graph period, 1880 to 1980, highly accurate metereological data for the entire area is available.

In the second graph the amount of data is much greater, and less controlled by local geographical variability. The temperatures shown (for the year and for June) are taken from the 1982 tabulations of Drs. P. D. Jones, T. M. L. Wigley and P. M. Kelly. The temperature scale is expanded. The temperatures shown are departures from a mean reference temperature for the period 1946 to 1960. In the early portion of the record the data covers about 20 percent of the northern hemisphere; toward the end it covers about 55 percent. This is not complete coverage, but it is enough to smooth out the annoying ups and downs visible on records of individual stations. On these curves the year-to-year wiggles are about 0.5°F.

There is some hope, then, that temperature drops due to eruptive events may be visible, even for volcanos less explosive than Mount Tambora. You should examine these temperature curves to see if you can detect drops in temperature following volcanos. Most meteorologists see drops following the large eruptions of Krakatoa, Santa Maria, Azul, Nilahue and Agung, but not Mount St. Helens. However, the correspondence is not perfect. What do you see? Is cause and effect proved beyond reasonable doubt, or is the relation merely suggestive? There is a commonsense standard of certainty; statisticians have a refined standard of certainty. But from data like this, they suffer agonies of indecision much as you are perhaps experiencing.

What do you think of the relation between annual sunspot numbers and the departure of temperature? Does the latter have a pronounced 11-to-12 year cycle? So far as you can see by inspection, do high sunspot numbers correspond to warm or cold periods? If for a moment you average over the year-to-year short-period fluctuations you can discern some long-period trends. From 1880 to 1940 northern hemisphere temperatures seem to have increased about 1°F, followed by a dip in 1960 to 1975. This is the slow kind of temperature change that climatologists talk about when they discuss possible effects on temperature due to long-term buildup of carbon dioxide in the air, resulting from increased burning of fossil fuels, such as coal and oil. Should the temperature departure on this graph someday exceed 3°F, they say, the corn belt might move to Canada.

The dashed line joining the peaks of the sunspot curves forms Eddy's "envelope." He points out that *it* parallels the long-term temperature changes. Do you think so?

Looking over these curves carefully, and trying to judge the truth of various hypotheses, is helpful in trying to understand the difficulty climatologists face in coming to conclusions—and why opinions vary. More sophisticated statistical tests of the data permit some quantification of the uncertainty—such as would be useful for setting odds for a wager—but they are not sufficient here to force definite conclusions. Intuition, upon examining this data, suggests that volcanos produce cold spells. We can use it as a working hypothesis, but not a proven fact. It also tells us that sunspots—if they have any effect at all—operate over longer periods of time.

CHAPTER 13

Judging Cause and Effect

A retrospective "impact statement"

THE QUESTION IS whether the great eruption of Mount Tambora in 1815 set off a series of events that caused the cold summer of 1816 and the social dislocation of following years. Did its influence extend to as late as 1832—the last year of the first great cholera pandemic?

Even without good volcanologic investigations of Mount Tambora, of global temperatures at so early a date, and a settled cause and effect theory about climatological phenomena, it seems possible to make a provisional environmental impact statement.

In summary, the cold summer of 1816 is explained as follows:

There was a general world-wide temperature lowering that persisted for several decades following the 1780 eruption of Mount Asama, possibly due to the volcanic dust, or perhaps even to the low level of sunspot activity.

During this time the random fluctuations of the internal redistribution of heat in the atmosphere due to large atmospheric eddies certainly did help produce local anomalies in temperature, and even of global mean temperature, which occasionally reached lower values than in times before or after. These are the result of the internal machinery of the atmosphere and not related directly to variations in the incoming solar radiation either by dust or sunspot activity. They introduce a random element in the records that complicates interpretation of the action of other causes.

The eruption of Mount Tambora in mid-1815 further depressed temperatures around the world in 1816, which helped make it stand out during the generally cool period, and mark it as an event that ranks as most extreme at many points where long records of temperature are available. During the decade 1811 to 1821 sunspot activity was unusually low, but during 1816 it was higher than any of the other years of the decade. This seems to rule out a particular depression of temperature by sunspots during 1816 itself.

Due to the peculiar link of events between New England and western Europe, discovered by Namias, a particular pattern of planetary waves became locked into the atmosphere over the North Atlantic Ocean during the summer of 1816.

This pattern steered a succession of cold waves down over New England and western Europe during the summer.

These cold waves depressed the mean temperatures locally in New England and western Europe more intensely than would otherwise have been the case.

Effects in America

The most critical effect was lowering the temperatures during the June, July and August cold waves enough to result in frosts that killed off the corn crop.

The abysmal growing weather of 1816 led to significant crop failures in a subsistence rural economy. The scarcity engraved itself on the minds of inland farmers who were most affected, and was retold to their descendants for the following century.

The extent of the crop failure in New England is somewhat arguable. But the fact of grain price increases is well documented, indeed they exceeded those of the succeeding years and were only finally overtaken by the rapid rise following the Russian wheat deal of 1972. Even granting partial failure of the staple corn crop in New England, overall American grain production was plentiful enough to supply large overseas shipments to Europe, where it was imported in unusually large amounts. In America this exportation of grain played an important role in raising domestic prices. In the commercial coastal towns of the northeast United States there was no prospect of famine—although as we have seen, there was a concern over availability of animal fodder.

The placid reports to the Philadelphia Society of Promotion of Agriculture describe the cold summer in measured terms: after all, a cold summer is more memorable to farmers than an occasional frigid winter that must be endured. The crops themselves depend on the summer. Corn flourishes best at the northern limit of cultivation. The harvest per acre in Connecticut for years far exceeded that of the famous mid-western "corn belt." Planted at its very northern limit in northern New England near the frost line, corn was vulnerable to the summer frosts of 1816. Isolated subsistence farmers on hard-scrabble hill farms had little to sell and little ready money with which to purchase supplies. Road transportation was so prohibitively expensive that they could not market bulk produce, such as corn. There were no investigative reporters to write up their plight, although there may well have been unscrupulous traders to sell them rotten mackerel. They were used to scrambling for food, even while their more prosperous fellow citizens were importing oysters from Albany and slaughtering their Christmas pork.

It was the largely unchronicled distress in these isolated, marginally viable hill farms that probably accounts for the legendary role of 1816 in New England folklore.

To the extent that 1816 drove families out of the hills to the western farmlands it might even be considered beneficial in the long run. That the cold summer actually drove significant numbers of settlers toward the west there can be little doubt.

Effects in Europe

In Europe the situation was much worse, the social disorganization following the end of the Napoleonic Wars compounded with the bad harvest, and true famine did ensue. And here we encounter a curious psychological phenomenon: although suffering much less, the Yankee farmers remembered 1816 much better than their European counterparts. It is possible that, viewed against the general domestic tranquility of the young United States, 1816 could stand out as something to remember; viewed against the turmoil of Europe, it would soon sink indistinguishably in men's minds into a confusion of greater disasters.

What about the link to disease? In the case of the typhus epidemic that began to spread from Ireland to the rest of the British Isles in late 1816, causing more than 65,000 deaths, there can be little doubt that it was related to the famine resulting from the anomalous summer weather.

Volcanos and cholera

Finally, we come to the question of the cholera pandemic. Until 1816, medical historians say cholera was restricted to India and sometimes extended as far as China, but not beyond. In 1816 there were bad harvests in India. The ensuing famine triggered a local epidemic in the densely populated countryside surrounding the moribund rivers above the mouth of the Ganges at Jessore in Bengal. But, unlike earlier occasions, this time it did not remain there long. It was spread by British military operations to Afghanistan and Nepal. Reaching the shores of the Caspian Sea, it slowly moved westward by two routes to the west: one up the Volga River to the Baltic ports and the other through the Moslem Hadj to the Middle East. In a time before airplanes and railways the progress of the cholera was leisurely. It did not survive long sea voyages.

But 16 years does seem an extraordinarily long time for the disease to reach Europe. It is likely the strain that ultimately reached Europe might have left India some years after the 1816 outbreak. In any case, the great cholera pandemic would surely have occurred a few years later, had not the Bengal drought of 1816 set it off. Europe was "opening up Asia" and in so doing would inevitably have drawn this pestilence from the Pandora's Box of empire.

Scientific information about the eruption of Mount Tambora and about world-wide distribution of the temperature anomalies in 1816 is too sparse to make this event primary evidence in proving the case for volcanic dust as a cause of climatic variation. On the other hand, more recent and better-documented temperature records of modern times show conclusively global temperature drops dramatically following some major eruptions.

Extrapolation of some recent numerical simulations suggests that Mount Tambora could have lowered global temperature by something like 1°C over a period of a year or two. This is comparable to large-scale trends of world temperature such as the slow temperature increase of 0.6°C in the northern hemisphere observed between 1857 and 1937.

Dust vs carbon dioxide

Continued scientific investigation of the effects of dust on climate, using numerical models of increasing sophistication, is extremely important to our civilization if we are to anticipate the possible effects of human intervention in the environment. The experience gained in trying to model the effects of dust from volcanos can be applied to calculating the effects of man-made dust and aerosols. Already we have some indications that such dust may actually be a help in offsetting the effects of carbon dioxide increases in the atmosphere that accompanies our continued burning of the world's coal reserves.

Without the dust to offset its effect, the carbon dioxide would increase world temperature to a point where large ice masses in the Antarctic might melt and raise sea levels all around the world—not to speak of disturbing agriculture everywhere.

As a final comment, it seems that the demonstration that Mount Tambora and other volcanos from time to time have given us of how the temperature of the world can be altered by an almost imperceptible amount of dust points to a mechanism that in the next century may save us all from sweltering under the pall of the carbon dioxide that will by then envelop us.

Within the next century the preservation of our climate—and indeed our survival—may depend upon a precarious balance of two poorly understood mechanisms, one tending to increase global temperature and the other to decrease it. That each of these mechanisms is strong enough to drive the climate beyond tolerable limits suggests how important it is to continue trying to understand the scientific basis of these processes. This will help us to gain the wisdom needed to manage rationally our beautiful world.

NOTES ON SOURCES

CHAPTER 1

One of the delights of preparing this book has been pursuing facts into a variety of back-alleys and dusty attics of history. There we found far too many curiosities and fascinating personalities to include in our work. Our search for material on the eruption of Mount Tambora was something of an exception. There is not much written on it.

Standard volcanological reference books referred us to a rare pamphlet by Heinrich Zollinger, 1855: *Besteigung des Vulkans Tamboro auf der Insel Sumbawa und Schilderung der Eruption desselben im Jahren 1815.* (Wintherthur, 1855, Zurcher and Fürber.) The only copy we could locate is in the British Museum Library. This work consists of two parts—an account of the author's ascent of the volcano years after the eruption and a German translation of an English language report by Sir Thomas Stamford Raffles to the Batavian Natural History Society in September, 1815: *Narrative of the effects . . . of the eruption from the Tambora . . . 11 and 12 April, 1815. (Verhandelingen van het Batavaaisch Genootsch. van Kunsten en Wetenschappen, deel* VIII, 2 editie, Batavia 1826, pp. 343.) We found a copy of this in the library of the Museum of Comparative Zoology at Harvard.

It might seem odd that a British military officer, governing the Dutch East Indies, should deliver a paper to a local Dutch scientific society. Raffles was, however, an odd man. Today his fame rests upon the founding of Singapore. He was born at sea July 6, 1781, at the height of the American Revolution in his father's West Indiaman *Ann,* four days out of Jamaica in a convoy of ships headed for England. He entered the employ of the East India Company at the age of 13 years, and 10 years later was assistant Secretary to the Council at the Presidency at Penang. (Collis, M., *Raffles.* Faber and Faber, London, 1966). Trevelyan writes of him as "one of the greatest and best servants our Empire ever possessed. He was perhaps the first European who successfully brought modern humanitarian and scientific methods" to improve the lot of the native races of Asia. (Trevelyan, G. M., *British History in the Nineteenth Century,* London, 1922.) Upon his death on July 5, 1826, his grave was left unmarked in the church at Hendon for 50 years due to the spite of the rector, who disapproved of Raffles' sentiments in favor of emancipating slaves. The rector owned a share in a west Indian plantation.

During his residence in Java, Raffles made the acquaintance of an American naturalist, Thomas Horsfield, M.D., who was living in the Indies. As a result, Raffles developed a deep interest in natural history and this doubtlessly stimulated his decision to document the events of the eruption. An account of the eruption is also given in his *History of Java;* London, 1817 (Oxford University Press, 1965).

It is from this source that Lyell obtained the material on Tambora for his famous book: Lyell, Sir Charles, 1830, *Principles of Geology* (Johnson Reprint Corp., 1969).

Aside from these various renditions of Raffles' account, we have found no other description of Mount Tambora's eruption.

Mount Tambora today is quiescent, except for some vents near the walls of the inside of the crater's rim from which gases are emitted. It has seldom been visited by scientists. Accounts of only three ascents (1847, 1913 and 1947) have been published. That by W. A. Petroeschevsky appeared in 1949 in the Dutch scientific journal, *Tijdschrift van het K. Nederlandsch Aardrijkskundig Genootschap:* Amsterdam Series 2, Vol. 66, pp. 688–703.

In case you wonder about what the Eleven Directions in the Great Rite celebrated on Mount Agung are, they are the eight points of the compass, up and down, and the center—the Holy Mountain itself.

CHAPTER 2

Anyone who seeks information about the history of American meteorological events can do no better than to turn first to the works of David Ludlum. We started our search with the short, but reference-filled, chapter on June 1816 in his *Early American Winters, 1604–1820* (1966, American Meteorological Society, Boston, Mass.), and pursued references and sources that he cited. Particularly valuable for anecdotal material was Sidney Perley's *Historic Storms of New England* (1891, The Salem Press, Salem, Mass.). We could also see, from Ludlum's example, that we could expect to find additional material in newspapers of the period, and spent happy days poring over the marvelous collection at the American Antiquarian Society in Worcester, Mass., the Massachusetts Historical Society in Boston, and the Provincial Archives in Halifax, Nova Scotia. The collection first named is particularly impressive, as it was founded upon the personal collection of Isaiah Thomas, a New England publisher who subscribed to almost all papers published at the time. It is interesting to note that these early newspapers were published on such good paper that one examines the originals themselves, whereas the crumbly newsprint of later decades is now available only on microfilm.

Ludlum's chapter also pointed the way to diaries of great interest. We especially value the Harwood diary (unpublished, at the Bennington Museum, Vermont) because it gives such a clear and detailed picture of activity and harvest on a well-managed farm during the whole period of the cold summer. It is difficult, when using a diary like Harwood's, to avoid becoming wholly engrossed by it. In a different manner the Reverend Thomas Robbins' diary fascinates the reader and easily leads him far from the cold summer.

Then there are those who, like Harvard's Samuel Williams (see Sibley's Harvard Graduates), whose role in meteorology was mainly in the years preceding 1816, but who is so personally colorful—he wore scarlet clothes in an otherwise black-suited faculty—that it is painful to limit ourselves to brief mention of his expulsion from his professorship. He mounted a solar-eclipse espedition to British-occupied Penobscot Bay during the Revolution. By mistakes in his calculations he set up his instruments in the wrong place, then covered it up in his report to the American Academy of Arts and Sciences by falsely claiming that the map of Maine was in error.

After much thought, we decided to omit reference to Williams' fascinating report to the same Academy on the Dark Day of May 19, 1780, when the sun disappeared without any eclipse at all, and the ingenious treatise on Climate Change in the Christian Era that he wrote from his refuge in Rutland, Vermont.

Likewise, Dr. Benjamin Waterhouse's measurements of July, 1816, temperature in Cambridge scarcely justifies recounting his feud with Harvard, which grew out of his disagreement with moving the medical school across the Charles River to Boston, resulting in his founding a rival medical school in Cambridge. In the case of Jeremiah Day, the Yale president whose meteorological record is crucial in documenting the extraordinary June 7, 1816, we were so fascinated by his retirement in youth from the ministry on account of a throat hemorrhage—allegedly brought on by "too strenuous preaching" —that we pursued the matter to the end of his long life of 94 years and even the autopsy report. But again, the issue seemed too tangential to the cold summer to include in this book.

The New Haven temperature record can be found published in several places. We have used that by A. Loomis *(The Meteorology of New Haven*, 1866, *Transactions of the Connecticut Academy of Sciences*, Vol. 1). Photographic copies of the other meteorological journals and temperature records quoted in this chapter can be obtained from the National Climatic Center in Asheville, North Carolina. Some of the originals have suffered a strange fate. Generations of New Englanders before 1958 regarded the Blue Hill Observatory in Milton, Massachusetts, as the font of all meteorological wisdom. Manuscript weather journals were bequeathed to it from time to time by families who inherited them. Charles F. Brooks, its director, was well beloved within the meteorological community, but not regarded highly by colleagues at Harvard who wanted to raise the intellectual level of meteorology as a science.

After Brooks died of a heart attack while shovelling coal in the cellar of his official residence (he had decided to spare the observatory the expense of an oil-burner), the collection of old journals was shipped to the National Climate Center, where they were photographed and have become a permanent part of the climate record. Friends tell of having retrieved the precious manuscript volumes themselves from the dust bin of the center.

Two very useful accounts of the cold summer have appeared in professional journals: the first is Willis I. Milham's *The Year 1816—The Causes of Abnormalities* (Monthly Weather Review, 1924, Dec., p. 563). The second is

Joseph B. Hoyt's *The Cold Summer of 1816* (Annals of the Association of American Geographers, 1958, pp. 118–131).

CHAPTER 3

Our reading of American newspapers indicated that 1816 was a bad year in Europe as well. A search of British meteorological literature confirmed it. For example, we found that Professor Gordon Manley, in a presidential address to the Royal Meteorological Society in London in 1946, had identified the summer of 1816 as something quite extraordinary *(Temperature Trend in Lancashire, 1753–1945,* Quarterly Journal of the Royal Meteorological Society, 1946, Vol. 72, p. 1).

Although June, 1816, was not extraordinary, July of 1816 was the coldest July average temperature in the entire 192 years of record. August, 1816, was not as extreme, there being 23 years as cold or colder. But on the 31st snow actually fell at Barnet, a short distance northwest of London. September 1816 was only 9th coldest in the 192 years of record.

Professor Manley's reconstruction of the temperature record in central England in the early 1800s depends upon a piecing together of records from many sources, as in America, largely from the journals of amateur English meteorologists. In the present day, when routine observations are obtained by various government bureaus, it is important not to underrate the competence of these old volunteer observers, nor their care and interest in acquiring good and accurate data. A Mr. Pitt of Carlisle can be offered as an example of one such amateur who contributed to the knowledge of this period. A description and summary of his efforts has been given by Thomas Barnes, M.D., physician to the Fever Hospital and Public Dispensary at Carlisle *(Transactions of the Royal Society of Edinburgh,* 1830, Vol. 11, p. 418 ff.):

> "For many years Mr. Pitt had no particular occupation and meteorology was his hobby . . . Mr. Pitt was seldom absent from home, and whenever any unavoidable circumstance obliged him to go a distance he always appointed a confidential person to take the observations for him (thermometer, barometer, rain gauge, appearance of the sky, and winds from the weather-cock of Carlisle cathedral). Mr. Pitt was in possession of several thermometers and barometers, which were of a superior kind, and he prided himself upon their goodness. The thermometer he used latterly was . . . constructed and graduated with great care, and has Reaumur's scale on one side, and Fahrenheit's on the other. It hangs upon the garden wall, in a glass cylinder, which is open at each extremity. It is not in contact with the wall, and is sheltered from the heavens, and the falling vapours. It is placed in a northwestern aspect, about six feet from the ground. A good location has been chosen for the instrument. The measurements were made thrice daily from January 1, 1801, through the end of December, 1824—not a single reading is missing."

According to the long records obtained at Geneva, the mean summer temperature of 1816 was the lowest for the entire period 1753–1960 *(Klimatologie der Schweiz,* 1961, Schweizerisch Met. Zentralanstalt, Teil 2, pp. C-20-23). Similar long records are available for many other European locations; this does not seem to be the place to list them, as they are available in standard climatological works.

The study of the dates of the wine-harvest by Charles Alfred Angot referred to is *Etude sur les vendanges en France* (Annales du Bureau central météorologique de France, 1883).

We discovered our first good references to actual famine in Europe in the public card index of the New York Public Library, and they form the basis of our remarks on the hunger in Switzerland and France: Jakob Keller-Höhn, *Die Hungernot im Kanton Zürich in den Jahren 1816–17* (Zürcher Taschenbuch Jhrg., 1948, 68, pp. 75–114); Robert Margolin, *Troubles provoqués en France par la disette de 1816–1817* (Revue d'Histoire Moderne, September 1929, pp. 423–460) and J. P. Housel, *Histoire des Paysanes français des XVIIIe siècle à nos jours* (edit. Howarth, 1976).

Subsequently we inquired of a friend in Germany whether he could locate some more material, and we were surprised to receive by return mail a copy of a splendid work by Professor J. D. Post: *A Study in Meteorological and Trade Cycle History: The Economic Crisis following the Napoleonic Wars* (The Journal of Economic History, 1974, pp. 315–349). This work, and a later book (J. D. Post, *The Last Great Subsistence crisis in the Western World;* Johns Hopkins Press, 1977), of which we became aware only in September, 1979, took us by surprise. Authors always run the risk that they may discover, after much research, someone else has already done the job they set out to do. After recovering from the shock, we decided that our works were not entirely parallel—but we did decide to abbreviate what little we had to say about Europe, and refer the reader to Post's works if he wants to read more—we do recommend it. There actually is a book with the title "The Year Without A Summer," but it is a children's book, mostly about a turtle.

Amongst those civilized countries of the early 19th century that kept some records of crop yield was Japan. Rice is a crop very sensitive to summer coldness, much like Indian corn. Our comments on the Japanese rice crops of 1816 are drawn from a paper by H. Arakawa *Meteorological Conditions of the Great Famines in the Last Half of the Tokugawa Period, Japan* (Papers in Meteorology and Geophysics, Vol. 6, 1955, pp. 101–115). Arakawa gives a chart of rice prices at Kyoto for the years 1773–1870. Rice during the years 1816 and 1817 does not appear to have been particularly high-priced. In fact the picture is dominated by immense peaks at 1787, 1837 and 1866, which he attributes to weather disturbances associated with the erruption of Mount Asama and Mount Coseguina (Nicaragua).

In the case of famines of 1787 following the Mount Asama eruption, the price of rice more than doubled and about 10 percent of the population of Tohoku—the northern portion of Honshu—perished, some persons being reduced to cannibalism. In the case of the 1837 famine the rice price rose even higher, and the crops were reduced to about 20 percent of the usual yield at the estates of the Sekiya family at Ryogo-Mura, Tochigi Prefecture, for example. These same records show nothing comparable in 1816–1817. During the great famines farmers tried to move—as they did in New England

—but in strict feudal Japan were counted in the census as absconders instead of as emigrants, a measure of ignominity that seems to have kept the numbers of refugees down to about 4 percent of the population.

CHAPTER 4

This chapter represents the results of a good deal of general reading about the history, economy, customs, demography and statistics of New York and New England in the first decades of the 1800s. The problem we faced was not how to find material, but how to distill and summarize it. It can serve no useful purpose to enumerate the many books, newspapers and diaries that we studied. Standard histories of the period list them. We did draw heavily upon Percy Bidwell's *Rural Economy in New England at the Beginning of the Nineteenth Century* (Transactions of the Connecticut Academy of Arts and Sciences, 1916, Vol. 20, pp. 241–399); William C. Langdon's *Everyday Things in American Life 1776–1876* (Scribner, 1941); and Paul W. Gates *The Farmer's Age* (Vol. III of The Economic History of the United States, Harvard University Press, 1960).

The travel journals of Timothy Dwight (*Travels Through New York and New England*, 4 Vols., New Haven, 1822; reprinted by Belknap Press, Harvard, Massachusetts, 1969) were helpful, especially in regard to the problems of road transportation, mostly for a period a decade or more earlier than 1816. David Thomas' description (*Travels Through the Western Country in the Summer of 1816*, Auburn, New York, 1819) coincides with the cold year, and very occasionally makes some comment about it. Washington Irving's portrait of a snug Dutch Hudson River farm in his *Legend of Sleepy Hollow* could be read to gain some sense of the disparity of development that could be found even within the bounds of New York State at the time. Other examples of the diversity of experience could be offered, such as the social life of Judge Van Ness, Miss Johnson and Lady Upton at Dr. Isaac Cooper's house on Lake Otsego all during the cold summer and ensuing winter. Though his neighbors are said to have survived on rotten mackerel, Dr. Cooper imported oysters from Albany. Even more in contrast to the pioneer-style life in the back country, was New York City, a cosmopolitan city even in 1816. Macdef Eden's farm was about to be transformed into Times Square. James Palmer (*Journal of Travels in the U.S. of North America*, London, 1818), who visited in 1817, wrote, "The things that struck me on my first walks in the city were the wooden houses, the smallness, but neatness of the churches, the colored people, the customs of smoking segars in the streets (even followed by some of the Children) and the number and nuisance of the pigs permitted to be at large—as to the rest it is much like a large English town."

The Englishman Henry Fearon wrote to friends in August, 1817, (*Sketches of America: A narrative of a journey of five thousand miles through the Eastern and Western States of America;* London, 1819):

"In Broadway and Wall St. trees are planted by the side of the pavement. Most of the streets are dirty.

"An evening stroll along Broadway, when the lamps are alight, will please more than one at noon-day. The shops look rather better . . . I disapprove most heartily of the obsequious servility of many London shopkeepers, but I am not prepared to go the length of those in N.Y. who stand with their hats on, or sit or lie along their counter smoking segars, and spitting in every direction, to a degree offensive to any man of decent feelings."

But we must digress no more.

CHAPTER 5

Various articles on corn can be found in old issues of the *Yearbook of the U.S. Department of Agriculture*. There are some bits of pleasant reading here and there about corn in *A Long, Deep Furrow*, (University Press of New England, Hanover, N.H.) by Howard S. Russell, 1976, who died at the age of 92 at Wayland, Massachusetts, in 1980.

CHAPTER 6

We are particularly indebted to David Ludlum for referring us to the investigation of the harvests of 1816 conducted by the Philadelphia Society for the Promotion of Agriculture.

The printed reports of the Philadelphia Society do not hint at famine. There was more concern for a sufficiency of animal fodder. But these reports are limited to coastal areas, and perhaps are not representative of the situation in the remote hill farms of northern New England. Unfortunately there were no investigative reporters, as there would be today, roaming the backcountry hills for a story. We did not find contemporary accounts of extreme hardship—certainly not of famine. However, later accounts suggest that the summer of 1816 was really quite difficult in the hills.

Charlton Ogburn (*Birch trees are graces of our wild forests*, Smithsonian, December 1977, p. 77) relates that in searching for sources of fodder for their sheep and cattle during the winter 1816–1817, Green Mountain farmers collected the inner bark of white birches and had it milled into flour. Mixed with animal fat, this kind of flour at one time was even used for human consumption in east Siberian famines.

CHAPTER 7

The price of food, no matter how vital to the consumer it may be, is inevitably dry reading. If we couple to this dryness the unfamiliarity of the value of money in earlier times, it is even more difficult to maintain one's interest. In 1816 a housemaid might expect to earn $10 a month, a naval captain $100. The total federal revenue was $35 million dollars, 1/20,000 that of today. The price for posting a letter might exceed that of a gallon of whiskey. It is easy to draw incorrect inferences from lists of prices alone. Therefore our best guide as to the effect of the cold summer of 1816 on food prices is to compare prices in years just before and after 1816.

When we first approached the problem of evaluating the price structure through the decades 1805 to 1825 we were confused by newspaper reports. The first coherent discussion that we encountered was that of Joseph Hoyt, which we have already mentioned above. Then we became aware of a useful monograph by Jodie Pierce: *Cultural Sensitivity to Environmental Change, 1816, The Year Without a Summer* (report from the Center for Climate Research, University of Wisconsin, 1974). This led us in time to David Ellis' *Landlords and Farmers in the Hudson-Mohawk Region 1790–1850* and the work by Anne Bezanson, Robert Gray and Miriam Hussey, *Wholesale Prices in Philadelphia, 1784–1861* (University of Pennsylvania, 1936).

A particularly useful source for economic data in the United States is *Historical Statistics of the United States, Colonial Times to 1970* (in two parts, bicentennial edition, U.S. Dept. of Commerce, 1975).

CHAPTER 8

The great romance of American history is the westward migration. It is this grand movement of people that we associate with the Oregon Trail, the Conestoga wagon and the Indian attack. It began even in colonial times when settlers penetrated into Kentucky and Tennessee. The 1816 emigration was only a part of it. We found the two following sources useful: L. D. Stillwell's *Migration from Vermont (1776–1860)* (Proc. Vermont Hist. Soc., 5(2), pp. 63–241) and Clarence A. Day's *A History of Maine Agriculture 1604–1860* (University of Maine Press, 1954).

Day's chapter on the westward migration is wonderful reading. Our other main source of information was the articles and letters in the 1816–1817 issues of the *Eastern Argus* of Portland, Maine.

CHAPTER 9

Many legends followed the "year without a summer" or "eighteen hundred and froze to death"—terms that in themselves are exaggerations and that give the impression of snow covering New England the entire summer. John Winchester's account of his uncle's chilly demise comes from a newspaper clipping on file at the Vermont Historical Society. It states that Aden Brush, who kept the stagecoach tavern in Cambridge in 1816, found the story in a scrapbook of his grandfather, Dr. Salmon Brush.

T. D. Seymour Bassett, in a short article on the cold summer of 1816 in the summer, 1973, issue of *New England Galaxy* (periodical of Old Sturbridge Village, Massachusetts) reprints a somewhat different version of James Winchester's reminiscence, and attributes the claim that farmer James Gooding killed all his cattle and then hung himself—presumably in desperation over the gloomy weather—to a clipping from the Concord, New Hampshire, *Monitor,* sometime in the 1890s. He discounts this story and also a tale of Anna Gilbert's father that Dorset, Vermont, farmers slaughtered thousands of sheep at Pea St. brook, as well as Bakerfield historian Elsie Well's story of how Nathaniel Foster saved his cornfield by cutting and burning pine night and day.

The wholesale slaughter of sheep may not by apocryphal, however. For example, P. W. Bidwell and J. I. Falconer (*History of Agriculture in the Northern United States 1620–1860*, Carnegie Inst. of Washington, 1925) quote contemporary references (e.g., *American Farmer IV*, 1822, p. 70) that tell of just such quixotic slaughter of very expensive flocks following the great slump in wool prices from $1.06 to 62¢ a pound in New York between January and October, 1816. Entire flocks of fine Merino sheep were given over to the knife. A butcher was offered a $1,000 ram for $1.

The story of Joseph Walker's toes seems to be true, at least if we can believe a contemporary newspaper account of this event that appeared in the September 11, 1816, Dartmouth, New Hampshire *Gazette:*

"In Peacham, Vermont, on the 7th of June, Mr. Joseph Walker, aged 88, lost himself in a wood in a snow storm, and his feet were frozen so that it was necessary to amputate his toes!"

Samuel Hopkins Adams undoubtedly wrote for effect. He wrote other stories (*Grandfather Stories*, Random House, 1955) about life on the pre-railroad Erie Canal days—including an improbable account of the 1832 Cholera epidemic in upstate New York, "The Monster of Epidemy." Our comment on his mention of wolves in the farmers' fields, establishing the fact that they indeed were a nuisance in New England (in Maine) as late as March, 1816, is from Cyrus Eaton's *Annals of Warren* (Masters and Livermore, Hallowell, Maine, 1877, p. 323).

The *Old Farmer's Almanac* for 1966 celebrated the 150th Anniversary of the Summer of 1816 with a double page of curious facts. Among them was the account of the ear of 1816 corn kept on exhibit on the wall of the Augusta Post Office—but the present postmaster has no recollection of it.

CHAPTER 10

References to the great Cholera Pandemic of 1816 to 1832 are to be found in William McNeill's *Plagues and Peoples* (Anchor Books, Doubleday, 1976), C. E. Rosenberg's *The Cholera Epidemic of 1832 in New York City* (Bulletin of the History of Medicine, 1959, pp. 37–46), R. E. McGrew's *The First Cholera Epidemic and Social History* (Bulletin of the History of Medicine, 1959, pp. 37–46), R. E. McGrew's *The First Cholera Epidemic and Social History* (Bulletin of the History of Medicine, 1960, pp. 61-ff), H. M. Madden's *The Cholera in Pest, 1831* (Bulletin of the History of Medicine, 1943, pp. 481–482), N. R. Barrett's *A Tribute to John Snow M.D.* (Bulletin of the History of Medicine, 1946, pp. 517–535), and Dr. Edward Rüppell, *Reise in Abyssinie.*

CHAPTER 11

From a meteorological point of view the most important material is records of air temperature made with mercurial thermometers, manufactured, standardized and calibrated according to exacting scientific standards. By 1816 there had been nearly a century of development of the meteorological thermometer. We found the discussion of the thermometer's development in W. E. Knowles Middleton, *A History of the Thermometer and its Use in Meteorology* (1966, Johns Hopkins Press) to be absorbing. From the account of the lovely spiral Florentine thermometer of Mariani made for the Duke of Tuscany and Christopher Wren's thermograph of the mid-17th century, through the increasingly sophisticated researches of the 18th century, the story of the thermometer is a fascinating one.

In only one respect were we disappointed—we have been unable to discover where most of the thermometers used by American observers in 1816 were manufactured. American makers became established later than 1816, and we can only assume that the thermometers used were made in Europe.

It is de rigeur to mention sunspots whenever climate is talked about. This tradition is at least 200 years old. We are abashed to admit that our only authority for the account of M. Ruoy's attempts to calm the chiliastic fever in Paris over the July appearance of spots on the sun in the Foreign Intelligence column of the *New Bedford Mercury*. But spots there were: The Reverend William Bentley of Salem recorded that on May 30, about one-half hour after sunrise, he saw and conversed about a large central spot upon the sun—seen through the morning haze (Diary of William Bentley, D.D., Salem 1905–14, 4 volumes), although he found the elephant that he had seen on exhibition in the town the day before much more astonishing. The astronomer at the observatory at Williams College told reporters that on June 7, "Nine groups of spots besides several single ones were scattered from the eastern to the western side of the sun. I could distinctly see between sixty and seventy, some of which were large and well defined" (New Brunswick, *Canada Royal Gazette*, July 30, 1816).

The most up-to-date tabulation of numbers of sunspots is by J. A. Eddy *Climate and the changing sun* (in *Climatic Change*, 1977, D. Reidel Publishing Co., Dordrecht—Holland, pp. 173-190). The tabulation shows that the decades between 1800 and 1820 were generally low in numbers of sunspots, but that 1816 lies at one of the peaks of the regular 11-year cycles. Those who hold that lack of sunspots favors cold will find it difficult to explain the fact that spotty 1816 was colder than the other years of its time. Those who hold that sunspots favor cold will find it puzzling that the generally low-sunspot decade was so much colder than later years. If one postulates a more complex relation perhaps an explanation can be achieved, at least Charles G. Abbott, in his 94th year, thought so (C. G. Abbott, 1966, Smithsonian Misc. Collection, 448, 7).

The reference to ideas about electrical heating within the Earth as an important factor in climate is from the New York Museum (1816, p. 99). Chladni's idea about icebergs in the Atlantic is in an article entitled *Ueber Ursachen des nasskalten Sommer von 1816 und zum Theil 1817* (Annalen d. Physik, 1819, pp. 132–136).

As an example of published sea-surface temperature measurements in the

North Atlantic, which show little deviation from what might be expected today, we mention the paper by John Hamilton, *Tables of Observations of the Winds, Currents, the Gulf Stream, the Comparative temperatures of the Air and Water, etc., made on the North Atlantic Ocean during twenty-six voyages to and from Europe (principally between Philadelphia and Liverpool) between the years 1799– 1817, inclusive* (Transactions of the American Philosophical Society in Volume II (new series), 1825, pp. 140–155). The speculations about large shifts in direction of the Gulf Stream as causes of the generally cold decades between 1780 and 1820 may be found in the following work: Reid A. Bryson and Thomas J. Murray, *Climates of Hunger* (The University of Wisconsin Press, 1977). We find their argument based on some old current charts engaging but unconvincing. Even today oceanographers have a great deal of difficulty understanding in detail how the ocean transports heat to Europe. Benjamin Franklin, the veritable founder of Gulf Stream and ocean thermometric studies, favored atmospheric dust as a cause of the cold winter of 1783–1784. Other works cited in this chapter are, in order of appearance: W. J. Humphreys, *The physics of the Air* (Lippincott, Philadelphia, 1920), S. H. Schneider and Clifford Mass, *Volcanic Dust, Sunspots, and Temperature Trends* (Science, Vol. 190, pp. 741–746), C. Mass and S. H. Schneider, *Statistical evidence on the influence of sunspots and volcanic dust on long-term, temperature records* (J. Atmos. Sci., 1977, 34, pp. 1995–2004), A. Robock, *Internally and Externally Caused Climate Change* (Journal of the Atmos. Sci., 1978, 35, pp. 1111–1122), H. E. Landsberg and J. M. Albert, *The summer of 1816 and Volcanism* (Weatherwise, 1974, pp. 63–66), R. A. Kerr, *Mt. St. Helens and a Climate Quandry* (Science, 211, January 23, 1981, pp. 371–374), C. V. Hammer, *Past volcanism revealed by Greenland Ice Sheet impurities* (Nature, 1977, 270, pp. 482–486), and C. V. Hammer, *Acidity of Polar Ice Cores in relation to absolute dating, past volcanism, and radio echoes* (J. Glaciology, 1980, 24, pp. 359–372).

CHAPTER 12

We hope that we have shown you how tantalizing the subject of climatology is. There are so many questions and doubts, and so few definite answers. If you care to dig deeper and would like to read a carefully presented independent empirical approach to the problems of volcanos and weather, based upon different tabulations of temperature and different estimates of the rank of eruptions, you can do no better than refer to the article by Dr. Robert C. Oliver in the September 1977 issue of the *Journal of Applied Meteorology*.

ACKNOWLEDGMENTS

We would like to thank Charles G. Bennett and Priscilla N. Kennedy, of the Bennington Museum, who were most generous with information about the Harwood Diary; to Elva Bogert, of the Massachusetts Historical Society, who helped us find our way through its extensive collections and to Irene Burkhardt, of Santa Barbara, California, owner of the German 1816 famine medal, who allowed it to be photographed.

We would like to make particular mention of the Massachusetts Horticultural Society, Boston; the Essex Institute in Salem, Massachusetts; the Old Dartmouth Historical Society, Dartmouth, Massachusetts; the Antiquarian Society, Worcester, Massachusetts; the American Meteorological Society, Boston; the Connecticut Historical Society; the New York State Historical Association; the Historical Society of Pennsylvania; the Vermont Historical Society; and the Filson Club, Louisville, Kentucky. Thanks are due the University of Maine Press for permission to quote from Clarence Day's *History of Maine Agriculture*. Access to the archived material in the National Climatic Center, Asheville, North Carolina, has been crucial to us.

Our biographical material comes from many sources; often we have depended heavily upon standard biographical works, in particular J. C. Poggendorf's *Biographisch-Literarischches Handwörterbuch* (Leipzig, 1863 ff), *Appleton's Cyclopedia of American Biography* (1889).

The editors of the *Old Farmer's Almanac* and of *The New Yorker Magazine* have helped us track down material in their publications.

We gratefully acknowledge the generous assistance offered by the college libraries of Bowdoin, Harvard (Museum of Comparative Zoology), Yale, Cornell and Middlebury; from the public libraries of Albany, New Bedford, Boston, New York and Plymouth (Massachusetts) and from the state libraries of Maine, New Hampshire and Vermont.

The following individuals have kindly helped us locate material outside the United States: J. F. Amirault and Gordon Riley, of Halifax, Nova Scotia; Nancy Simmons, Geneva; Mary Swallow, Great Britain; Paul Tchernia and Michele Fieux, Paris; and Walter Zenk, Kiel, West Germany.

The photographs of the caldera of Mount Tambora were supplied by Dr. J. D. Foden of the University of Adelaide, Australia, who participated in the 1976 visit, and by Dr. Adjat Sudradjat, director of the Indonesian Volcanological Section of the Department of Mines and Energy.

And finally, we wish to thank friends who read through portions of early drafts of this work and offered their advice: C. Godfrey Day, Kenneth Heuer, John T. Hough, R. B. Montgomery, Jerome Namais and A. C. Redfield. George Cresswell and Stephen Self were helpful in locating material on Mount Tambora. And most particularly we would like to thank our editor, Jim Gilbert, for helping us bring order to our manuscript.

INDEX